# ELECTRICAL INSTALLATION COMPETENCES

# Part 2 Studies: Theory

## Maurice Lewis
BEd (Hons), FIEIE

Stanley Thornes (Publishers) Ltd

First published in 1995 by:
Stanley Thornes (Publishers) Ltd
Ellenborough House
Wellington Street
CHELTENHAM GL50 1YD
United Kingdom

A catalogue record for this book is available from the British Library.

ISBN 0 7487 1661 0

Typeset by Florencetype, Stoodleigh, Tiverton, Devon
Printed and bound in Great Britain by The Bath Press, Avon

# Contents

# Preface

This is the final book in Part II studies concerned with the City and Guilds 2360–8 course on electrical installation competences. The book not only covers topics found in the existing 2360 scheme but also provides coverage of material found in the newly accepted lead body standards for NVQ Level III. This is the award required by a registered electrician in the electrical installation industry.

Chapter 1 deals with current legislation, summarising many statutory and regulatory requirements associated with electrical safety. It also describes procedures for forming a contract and explains the roles of important electrical organisations within the contracting industry. Site procedures for the electrical contractor are also discussed.

Chapter 2 is concerned with selecting and installing wiring systems and extends students' knowledge gained from reading the author's Part I Theory book. This chapter looks in depth at several common wiring systems in terms of their selection, erection and installation. Fibre optic cables are introduced and numerous examples are provided in Exercise 2 for students to complete.

Chapter 3 covers the broad area of inspecting, testing and commissioning installations and highlights some of the important considerations when dealing with emergency lighting systems, fire alarm systems and petrol filling stations.

Chapter 4 highlights further considerations for installations dealing with baths and showers, swimming pools, hot air saunas, construction sites, agricultural and horticultural premises, caravan sites, highway services. It covers protection for safety, restrictive conductor locations, equipment having high earth leakage and street furniture.

Chapter 5 provides the opportunity for students to revise on topic areas relating to other chapters as well as complete additional written exercise questions.

Appendices 1 to 3 provide the opportunity to study client/architect relations and also study forms on planning permission and contract of employment. Appendix 4 is devoted to answers in the chapter exercises.

It is expected that students will make reference to listed documents, particularly the author's other books in the series. Where relevant, the latest amendments of the *IEE Wiring Regulations* have been incorporated to cover the IEE *Requirements for Electrical Installation* effective January 1995.

*Maurice Lewis*

# Acknowledgements

The author wishes to thank the following manufacturers and organisations who have contributed information to make this book possible:

Amalgamated Engineering and Electrical Trade Union
British Standards Institution
City and Guilds of London Institute
Electrical Contractors' Association
Electricity Training Association
Health and Safety Executive
Institution of Electrical Engineers
Joint Industry Board for the Electrical Contracting Industry
J.T. Limited
Marshall Tufflex
National Inspection Council for Electrical Installation Contracting
RIBA Publications Limited
The NTN Partnership Engineering Services

# Communications and industrial studies

**1**

---

## Objectives

After working through this chapter you should be able to:

1 *state the main purpose of the following statutory and regulatory documents:*

- *Health and Safety at Work etc. Act, 1974*
- *The Electricity Supply Regulations, 1988*
- *The Control of Substances Hazardous to Health Regulations, 1988*
- *The Electricity at Work Regulations, 1989*
- *The Construction (Head Protection) Regulations, 1989*
- *The Low Voltage Electrical Equipment (Safety) Regulations, 1989*
- *The IEE Wiring Regulations, 1992 (BS 7671)*
- *The Building Regulations, 1984;*

2 *describe the procedure and documentation in forming a contract;*

3 *state the use of different types of working drawings, such as block plan, site plan, location and component drawings;*

4 *state the role and object of the following electrical organisations:*
- *Electrical Contractors Association*
- *Joint Industry Board for the Electrical Contracting Industry*
- *Amalgamated Engineering and Electrical Union*
- *National Inspection Council for Electrical Installation Contracting*
- *J.T. Limited*
- *Electricity Training Association;*

5 *describe site procedures for an electrical contractor concerning site organisation and administration of work.*

## Statutory and regulatory requirements

### *Acts and statutory instruments*

The legislation that governs people in Britain becomes law through an Act of Parliament. This process starts as a parliamentary Bill passing through the House of Commons, the House of Lords and then receives Royal Assent. To ease this burden of law making, Parliament can delegate some of its powers by passing Acts that only outline the principles of the law. In this way, ministers, government departments and local authorities can make their own rules and orders. These regulations are known as statutory instruments.

Five fairly recent statutory instruments affecting people in the electrical installation contracting industry are as follows.

- 1988 No. 1057: Electricity – The Electricity Supply Regulations 1988
- 1988 No. 1657: Health and Safety – The Control of Substances Hazardous to Health Regulations 1988
- 1989 Np. 635: Health and Safety – The Electricity at Work Regulations 1989
- 1989 No. 2209: Health and Safety – The Construction (Head Protection) Regulations 1989
- 1989 No. 728: Consumer Protection – The Low Voltage Electrical Equipment (Safety) Regulations 1989

The Act responsible for the emergence of many present-day statutory instruments is the *Health and Safety at Work etc. Act 1974* (HSW Act). It was the result of recommendations made by a Royal Commission in 1970 which looked at the whole field of health and safety at work. The purpose of the Act is to provide the legislative framework to promote, stimulate and encourage high standards of health and safety at work. This is achieved through the two bodies set up under the Act known as the Health and Safety Commission (HSC) and the Health and Safety Executive (HSE).

One of HSCs main functions is to submit proposals to the appropriate government minister for new regulations or approved codes of practice. Most of these proposals go to the Secretary of State for Employment after consultation with industrial groups such as the Trade Union Congress (TUC), the Confederation of British

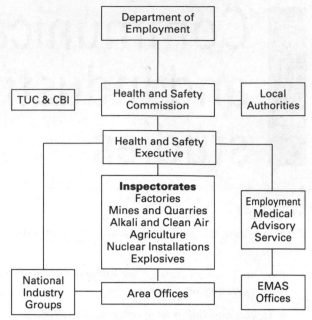

Figure 1.1   Organisations involved in the administration of statutory and regulatory requirements

Industry (CBI) and also local authority associations (see Figure 1.1).

### *The Electricity Supply Regulations 1988 (ESR)*

This was introduced by the Secretaries of State for Energy and for Scotland and replaced similar regulations made in 1937 and the Electricity (Overhead Lines) Regulations 1970. The new Regulations impose requirements on suppliers of electricity (the twelve regional electricity companies) regarding the installation and use of their electric lines and apparatus. Part IV of the ESR concerns the supply of electricity to consumers' premises. Of importance to designers of electrical installations is Regulations 30 and 31. Regulation 30 is concerned with a supplier's declaration of phases, frequency and voltage at the supply terminals. Regulation 31 is concerned with information designers may request from a supplier, such as the maximum prospective short circuit current, maximum external earth loop impedance, and type and rating of the supplier's cut-out fuse. Figure 1.2 illustrates these points in a TN–C–S earthing system.

### *The Control of Substances Hazardous to Health Regulations 1988 (COSHH)*

This was introduced by the Secretary of State for Employment. It called for a new legal framework

2

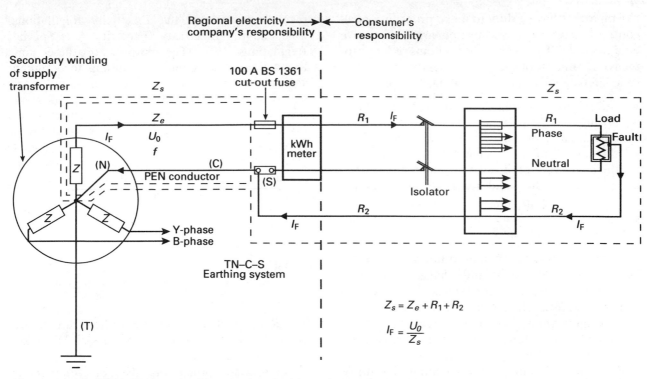

**Figure 1.2    Information required by an electrical installation designer**
Key:   $Z_s$ earth fault loop impedance path
$Z_e$ external earth fault loop impedance path
$R_1$ phase conductor resistance
$R_2$ protective conductor resistance
$U_0$ nominal voltage of the supply = 230 V +10% −6%
$I_F$ earth fault current flowing round the earth loop
$f$ supply frequency = 50 Hz ± 1%

$$Z_s = Z_e + R_1 + R_2$$

$$I_F = \frac{U_0}{Z_s}$$

for controlling people's exposure to hazardous substances in the workplace. Regulation 6 requires employers to assess all risks to health from work with hazardous substances. These are:

- substances listed under the *Classification, Packaging and Labelling Regulations 1984* as very toxic, toxic, harmful, corrosive or irritant;
- any substance for which a maximum exposure limit (MEL) or occupational exposure standard (OES) has been set;
- substantial quantities of any dust;
- micro-organisms;
- any other substance creating a comparable hazard.

Some hazardous substances found on site are:

*cement and mortar* (causing dermatitis, eye, mouth and nose irritation);
*wood dust* (causing breathing problems and eye irritation);
*man-made fibre (mineral wool)* (causing breath-

ing, eye and skin irritation);
*fumes/gases* (cause asphyxiation and may be very toxic);
*solvents* (glues, thinners etc. are harmful, entering the body through inhalation, and causing contact dermatitis);
*resins* (causing asthma and breathing irritation);
pesticides (causing skin irritation);
*lubricants* (causing dermatitis, acne and even cancer);
*acids* (causing burns and breathing irritation).

The COSHH Regulations make it a duty for every employer to inform, instruct and train his employees if hazardous materials are used on site. A main building contractor on site has a duty not only to his own employees but also to others working on the site. He should make sure that subcontractors and self-employed people have made their own COSHH assessments and have made effective arrangements to ensure that precautions are taken and supervised.

3

Employees have a duty to make proper use of all control measures including **personal protective equipment** (PPE) which includes items such as protective clothing, footware, eye protector and respirators. They should be given the following information:

- the nature and degree of the risks to health arising as a consequence to exposure, including any factor(s) that may influence that risk;
- the control measures adopted, the reasons for the measures and how to use them properly;
- the reasons for, and when to use, personal protective equipment;
- the procedures for monitoring hazardous substances and access to records.

In terms of instruction and training, employees should know what to do and what precautions to take, so as not to endanger themselves and others through exposure to hazardous substances. They should know what cleaning, storage and disposal procedures are required and when to carry them out as well as know emergency procedures.

Employers should start by making a suitable assessment of the risks and take the necessary precautions required before any work is carried out. Five simple steps for completing COSHH assessment are as follows.

1. *Gather information about the substances, the work and working practices* (e.g. what substances are present, how hazardous are they and who is likely to be exposed).
2. *Evaluate the risks to health* (e.g. how often is exposure likely to occur and what is the level of exposure).
3. *Decide what to do* (e. g. prevent or control the exposure, provide information and instruction).
4. *Keep a record of the assessment* (e.g. list the name, ingredients, supplier, hazard and storage of each substance).
5. *Regularly review the assessment* (e.g. are the risks serious and/or likely to change with changing circumstances).

**Note:** The HSE has produced a number of publications on this topic. One guidance booklet which is relevant to your work is: *The control of substances hazardous to health in the construction industry.*

*The Electricity at Work Regulations 1989 (EWR)*

This was made by the Secretary of State for Employment. It revoked many early orders and regulations such as the Electricity Regulations 1908 and the Electricity (Factories Act) Special Regulations 1944. The purpose of these new Regulations is to require precautions to be taken against the risk of death or personal injury from electricity in work activities. It applies to all work situations and all electrical equipment. Everyone at work, employers, employees and the self-employed have to comply with the regulations. Parts I and II of the Regulations comprise sixteen regulations that are of particular importance to people working on electrical systems, which includes all electrical equipment. (Figure 1.2 shows a TN–C–S earthing system).

Regulation 4(1) states:

All systems shall at all times be of such construction as to prevent, so far as is reasonably practicable, danger.

The Regulation does not specify how this should be achieved. The word 'construction' has wide application and hence the qualifying phrase 'so far as is reasonably practicable'. The *Memorandum of Guidance on the EWR (HS (R) 25)*, points out that it will include all aspects of the system's design. For example, electrical equipment should only be used for the purpose for which it is intended and in the environment for which it was designed and constructed (see Figure 1.3).

To comply with this regulation means that the system needs to be properly designed and maintained and the connected equipment needs to be regularly inspected and tested throughout the life of the installation.

A guide to the intervals between periodic in-

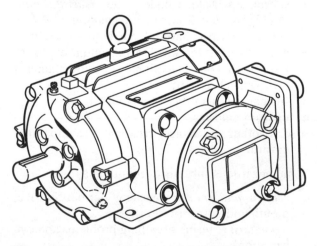

Figure 1.3 Flameproof motor for Zone 1 hazardous areas

spection and testing of installations is shown in Table 1.1.

Table 1.1 Periodic inspection and testing for different types of premises

| Type of installation | Maximum period between inspections |
|---|---|
| **General:** | |
| Domestic | 10 years |
| Commercial | 5 years |
| Educational establishments | 5 years |
| Hospitals | 5 years |
| Industrial | 3 years |
| | |
| **Buildings open to the public:** | |
| Cinemas | 1 year |
| Churches, under 5 years old | 2 years |
| Churches, over 5 years old | 1 year |
| Leisure complexes | 1 year |
| Places of public entertainment | 1 year |
| Restaurants and hotels | 1 year |
| Theatres | 1 year |
| | |
| **External installations:** | |
| Agricultural and horticultural | 3 years |
| Caravans | 3 years |
| Caravan sites | 1 year |
| Highway power supplies | 6 years |
| | |
| **Special installations:** | |
| Emergency lighting | 3 years |
| Fire alarms | 1 year |
| Laundrettes | 1 year |
| Petrol filling stations | 1 year |

Another regulation worthy of comment is Regulation 16 which requires persons to be competent to prevent danger and injury. It states:

No person shall be engaged in any work activity where technical knowledge or experience is necessary to prevent danger or, where appropriate, injury, unless he possesses such knowledge or experience, or is under such degree of supervision as may be appropriate having regard to the nature of the work.

This is an absolute regulation (see Note 1 below) to ensure that persons are not placed at risk due to a lack of skills on the part of themselves or others in dealing with electrical equipment. The regulation does not impose any requirement concerning age, qualifications or experience of persons supervising the work. It is the responsibility of an employer (dutyholder) to determine what is appropriate. In defence of an alleged breach of this regulation, it would be necessary for an employer to show that appropriate training had been provided.

**Note 1:** Regulation requirements that contain the term 'so far as is reasonably practicable' mean that the degree of risk has to be balanced against the time, trouble, cost and physical difficulty of taking measures to avoid the risk. If a person is prosecuted for failing to comply with a regulation requirement incorporating these words, the person may have to prove to a court of law that there was no better practicable way of meeting the requirement. Requirements that do not contain the above term are regarded as 'absolute' and must be met regardless of cost or any other consideration.

**Note 2:** Failure to comply with a regulation is a criminal offence under Section 2–6 of the HSW Act. Fines in a Magistrates Court can be up to £20,000 and a prison sentence of up to six months can be imposed for breach of an improvement or prohibition notice or court order.

**Note 3:** You will find a mention of the HSW Act and some listed statutory instruments (SIs) in the author's *Part 1 Studies: Theory* and *Electrical Installation Technology 3*. For more information on the HSW Act itself and explanation of terms used in Note 1, you should make reference to:

- HSC guidance literature – *A guide to the Health and Safety at Work etc. Act 1974*.
- HSE booklet HS (R) 25 – *Memorandum of guidance on the Electricity at Work Regulations 1989*.

*The Construction (Head Protection) Regulations 1989*

This statutory instrument came into force in 1990 and requires suitable head protection to be worn at all times during construction work unless there is no risk of injury from falling objects or a person hitting their head against something. On a construction site a safety helmet will need to be worn during most construction work.

**Note:** More about these Regulations can be found in chapter 2, page 18 of the author's *Part 1 Studies: Theory*.

*The Low Voltage Electrical Equipment (Safety) Regulations 1989*

These regulations came into force on the 1st June 1989 and are concerned with consumer protection. Regulation 5 states the requirements for electrical equipment to be safe and constructed in accordance with principles generally accepted within the member states of the European Community. Electrical equipment which satisfies the safety provisions of harmonised standards shall be taken to satisfy this requirement.

Figure 1.3 shows a flameproof motor which is used in a Zone 1 hazardous area. To meet the requirements of these Regulations it must withstand an internal explosion of flammable gas or vapour which may enter it, without suffering any damage to itself. It must not be capable of transmitting the internal explosion into the hostile environment through its joints and structural openings.

**Note:** You should make reference to the preface page of the *IEE Wiring Regulations* which shows a list of CENELEC harmonised documents incorporated in various parts of the Regulations.

*Other regulatory requirements*

These generally take the form of HSE memoranda, guidance notes (see Note 1 below), approved codes of practice, British Standards, BS codes of practice, safety training literature, etc. A selection of these documents follow.

- HSE HS (R) 25: Memorandum of Guidance on the Electricity at Work Regulations 1989
- HSC 1981/917: Approved Code of Practice Health and Safety (First-Aid) Regulations 1981
- BS 7671: 1992: Requirements for Electrical Installations IEE Wiring Regulations Sixteenth Edition
- HSE Guidance Note GS 27: Protection against electric shock 1984
- HSE Noise Guide No. 1: Legal duties of employers to prevent damage to hearing 1990
- IEE Guidance Notes No. 4: Guidance note on protection against fire 1992

These non-statutory safety guides are not intended to be an authoritative interpretation of the law but act as guidance information for dutyholders and others.

The HSE publish a wide range of safety information to support its legislation. For example, the purpose of the EWR Memorandum (HS (R) 25) is to give detailed explanation of the regulations' requirements. Approved codes of practice, such as the *Safety Representatives and Safety Committees 1988* and *British Standard Specification for Rubber Gloves for Electrical Purposes (BS 697: 1986)* are issued under Section 16 of HSW Act 1974 (see Note 2 below).

All British Standards are produced by the British Standards Institution (BSI). Although these are mainly technical agreements laying down requirements for materials, products and processes, HSE give support to a number of their safety publications. A typical example is the *16th Edition of the IEE Wiring Regulations (BS 7671:1992)* (see Note 3). These are non- statutory regulations but achieve compliance with the *Electricity at Work Regulations 1989*. Regulation requirements in Chapter 13 of BS 7671:1992 can be compared with EWR regulations 4–15.

The main reason for the conversion of the *IEE Wiring Regulations* into BS7671 was to allow electrical contractors in Britain to comply with the European Community (EC) utilities and public contracts directives. This requires a national standard to be quoted for wiring installations in large contracts.

To assist in the understanding of the IEE Wiring Regulations, a number of guidance notes have been published. These are as follows.

- *On-Site Guide*
- *No. 1 Selection and Erection*
- *No.2 Isolation and Switching*
- *No.3 Inspection and Testing*
- *No.4 Protection Against Fire*
- *No.5 Protection Against Electric Shock*
- *No.6 Protection Against Overcurrent*

These should be read in conjunction with the following statutory regulations:

1988 No. 1057: Electricity – The Electricity Supply Regulations 1988
1988 No. 1657: Health and Safety – The Control of Substances Hazardous to Health Regulations 1988
1989 No. 635: Health and Safety – The Electricity at Work Regulations 1989
1989 No. 2209: Health and Safety – The Construction (Head Protection) Regulations 1989

**Note 1:** There are currently five series of guidance notes available from the HSE, namely:

CS (Chemical Safety)
EH (Environmental Hygiene)
GS (General Series)
MS (Medical Series)
PM (Plant and Machinery)

Examples:

- *CS1: Industrial use of flammable gas detectors, 1987*
- *EH63: Vinyl chloride: toxic hazards and precautions, 1992*
- *GS47: Safety of electrical distribution systems on factory premises, 1991*
- *MS24: Health surveillance of occupational skin disease, 1991*
- *PM55: Safe working with overhead travelling cranes, 1985*

**Note 2:** The HSC has the power to issue approved codes of practice that provide practical guidance on how to comply with the legal requirements of the Act. While no one can be prosecuted for failing to follow the guidance of an approved code of practice, a defendant may have to prove to a court of law that he/she complied with the requirements of the Act in another way.

**Note 3:** The 16th Edition of the IEE Wiring Regulations, Requirements for Electrical Installations will be referred to throughout this book as the Wiring Regulations.

# EXERCISE 1.1

1. State the scope and purpose of the following statutory instruments made under the HSW Act 1974.
   (i)   *Noise at Work Regulations, 1989*
   (ii)  *The Health and Safety (First-Aid) Regulations, 1981*
   (iii) *The Health and Safety Information for Employees Regulations, 1989*

2. State the purpose of the following HSE guidance notes.
   (i)   *GS 6 (Rev) Avoidance of danger from overhead electrical lines, 1991*
   (ii)  *GS 31 Safe use of ladders, step ladders and trestles, 1984*
   (iii) *GS 38 (Rev) Electrical test equipment for use by electricians, 1991*

3. State the purpose of the following IEE/BSI publications.
   (i)   *BS 7671 IEE Wiring Regulations 1992*
   (ii)  BS 3456/IEC 335 Safety of household and similar electrical appliances, general and particular requirements, 1990
   (iii) *IEE Guidance Notes No. 5 on protection against electric shock, 1992*

## General requirements for buildings

If a person intends to build a new building or alter an existing building or put an existing building to a different use, building regulations approval will probably apply. These regulations were made under the *Building Act, 1984*. Their purpose is to ensure that buildings are safely constructed for the people who occupy them. Some buildings are exempt from this control such as buildings covered by other legislation and those not used by people. Also excluded are agricultural buildings, temporary buildings, glasshouses, mobile homes, small detached buildings and extensions up to 30 m² of floor area such as conservatories and porches.

In practice, if building regulations apply, the local authority needs to be notified. If planning permission is required, the person who requires the work done should submit a planning application. This is to give the local authority and public the opportunity to consider if the proposed building is in the general interest of the locality. Application is made to the planning department of the local authority. It may take the form of outline planning permission or full planning permission. The former is mainly to obtain approval of a proposed development without the obligation and costs of providing full working drawings. The latter require several copies of standard forms, plans and drawings (see Appendix 1).

### Contract procedures

In practice, the first step in a building project is for a prospective client to appoint an architect. He will act for him in the construction or alteration of the proposed building. On being appointed, the architect will obtain from the client, a brief, consisting of full details for him to produce sketch plans, etc. These stages are discussed in Appendix 1.

The architect will then apply to the local authority for outline planning permission. When approval is obtained, a design team comprising the architect, structural engineer, service engineer and a quantity surveyor will meet and discuss the brief. They will produce working drawings, specifications, schedules and bills of quantities. Full planning permission and building regulations approval will then be applied for to the local authority.

When these approvals are obtained, contract documents will be prepared and sent out to several building contractors for them to price and submit tenders (see Note 1 below). The returned tenders are then considered by the quantity surveyor who will advise the architect and client of the most suitable contractor. The client and contractor will then sign the contract. When the contract is signed the bills of quantities and prices become part of the contract agreement and will be used in the preparation of the final account and in the settlement of variations. (See the terms used in contract clauses under the subheading 'conditions of contract'.)

The provision of bills of quantities leads to accurate tendering, as all the tenders have identical documents from which to work. Contract documents normally consist of working drawings, bills of quantities, specifications, schedules and conditions of the contract. The procedure stages are shown in Figure 1.4.

**Note 1:** The term 'tendering' is used to explain the way a person (normally a contractor) obtains work via the preparation and submission of a tender (i.e. a formal offer to carry out work at a stated price). The object is to submit a price which covers the cost of carrying out proposed work and provide a profit. The term must not be confused with 'estimating' which is the technical process of predicting costs such as component costs, installation costs and overhead costs. This entails studying contract documents such as drawings, specifications, etc. in order to identify the materials and labour necessary.

*Working drawings*

These drawings are often prepared under the guidance of BS 1192 so that information needed for the proposed building is expressed accurately. They will include a block plan, site plan, general location drawings and sometimes component drawings.

The **block plan** is normally to a scale of 1:2500 or 1:1250 and identifies the site. It also locates the outline of the proposed building in relation to a town plan or wider area (see Figure 1.5).

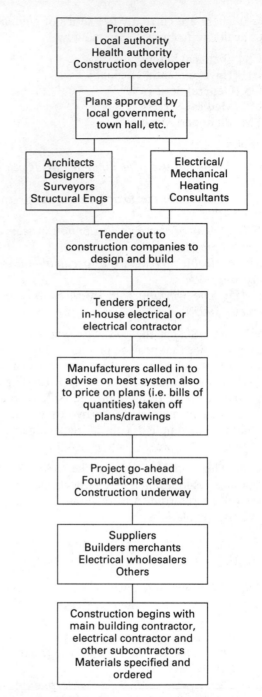

Figure 1.4 Building project development and procedure

The **site plan** is normally to a scale of 1:500 or 1:200 and identifies the position of the proposed building for setting out purposes. It also gives a general layout of the site showing services, drainage and landscaping (see Figure 1.6).

General **location drawings** are normally to scales of 1:200, 1:100 or 1:50 and show the position of the

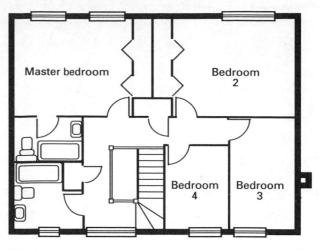

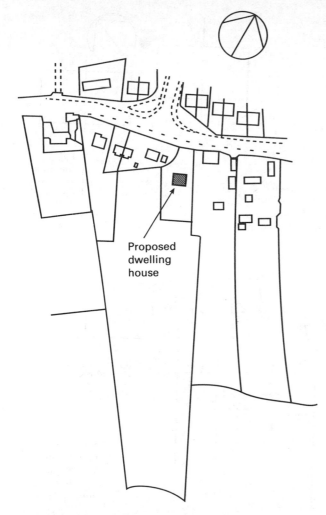

Figure 1.5   Block plan (scale 1:2500)

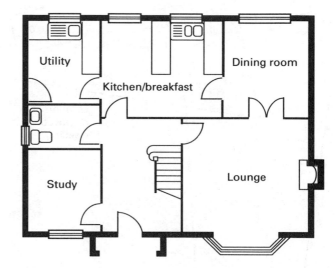

First floor

Ground floor

Figure 1.7   General location drawing (scale 1:100)

various spaces in the proposed building such as the hall, lounge, bedroom, etc. (see Figure 1.7). Other drawings, such as assembly drawings and component drawings show in detail the constructional elements of the proposed building and are produced to scales of 1:20, 1:10 or 1:5.

For showing accurate detail, orthographic and isometric drawings are used. The **orthographic drawing** shows a separate drawing of all the views of an object (its plan, front elevation and end elevation) on the same drawing sheet. These drawings can either be expressed in first angle projection (used for building work) or in third angle projection (used for engineering work). The **isometric drawing** is a pictorial drawing of an object without showing its sectional views. All horizontal lines are drawn at 30° to the horizontal and all vertical lines are drawn vertically (see Figure 1.8).

Electrical drawings are mainly **layout drawings** produced in ink on paper negatives. The architect or consulting engineer will provide the electrical contractor with the proposed building outline for laying out his services. The drawings will show the actual position of the electrical equipment. If an architect's instruction (AI) alters the installation then the layout drawings will need to be revised.

Figure 1.9 shows a typical layout drawing of the electrical equipment required for a light engineering workshop. The scale is 1:50 meaning that

9

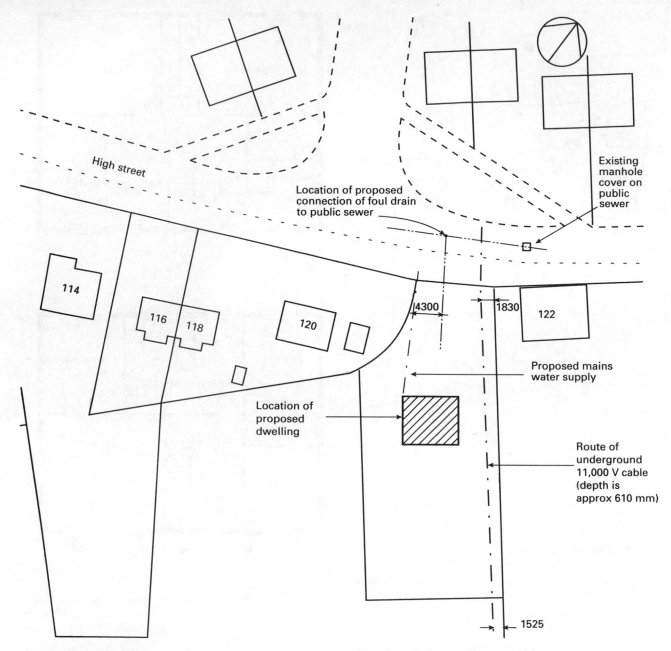

High street

Location of proposed
connection of foul drain
to public sewer

Existing
manhole
cover on
public
sewer

114

116  118

120

4300   1830   122

Location of
proposed
dwelling

Proposed mains
water supply

Route of
underground
11,000 V cable
(depth is
approx 610 mm)

1525

**Figure 1.6**  Site plan (scale 1:500)

the drawing is one fiftieth of the actual size it represents. By placing a metric ruler over the drawing, the length of the workshop can be determined as 10.25 m (1 cm = 50 cm or 0.5 m).

When electrical work on site progresses, layout drawings can be modified to keep a record of the work done. They can be used to show the routes taken by the installed wiring systems, the method of installation and the final position of all the fixed

equipment. When the drawings are used for this purpose they are called 'as fitted' drawings and must be kept up-to-date. On completion of the project copies of the drawings are handed over to the client or owner of the building. Other electrical drawings that you are likely to come across are block diagrams, circuit diagrams and schematic diagrams.

**Block diagrams** show electrical equipment as

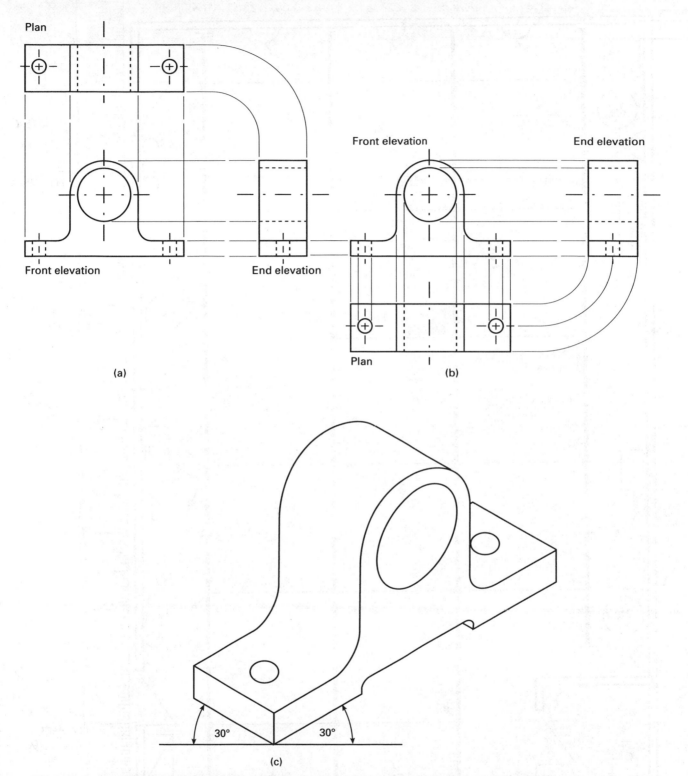

Plan

Front elevation

End elevation

(a)

Front elevation

End elevation

Plan

(b)

30°        30°

(c)

Figure 1.8   Orthographic projection and pictorial views of a pedestal bearing: (a) 3rd angle projection; (b) 1st angle projection; (c) isometric view

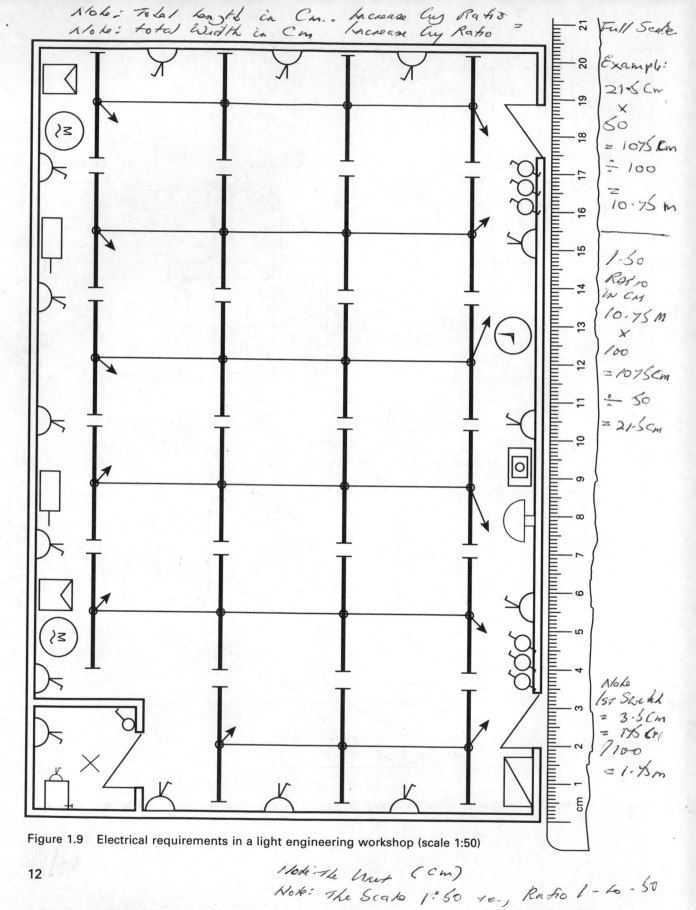

Figure 1.9    Electrical requirements in a light engineering workshop (scale 1:50)

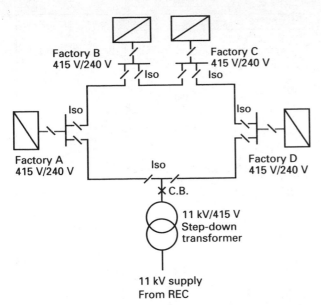

Figure 1.10 Line diagram of a ring main distribution system feeding separate factory premises on one site

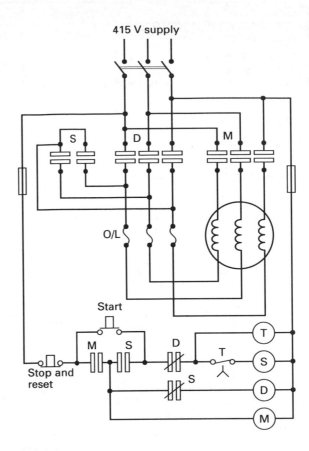

Figure 1.11 Schematic diagram of an automatic star/delta starter

blocks connected together with a single line. It allows the diagram to be easily understood and avoids showing the internal connections within the equipment. Figure 1.10 shows a **line diagram**.

Circuit diagrams show the destination of circuit conductors to the components that they connect. There are many examples of such circuits in this book and it is very important to label the correct polarity of circuit conductors. Where possible the circuits should be drawn using **BS 3939 circuit diagram symbols**, not graphical location symbols. For complex circuits, especially those having several operating sequences, it is preferable to use schematic diagrams (see Figure 1.11).

### Bills of quantities

These are prepared by a building quantity surveyor and give a complete description and measure of materials, labour and other items required to carry out the proposed work based on all the drawings, specifications and schedules. The bill of quantities is a method of tendering (see Figure 1.12) and ensures that all estimators prepare their tender on the same information. It avoids measuring errors or omissions on the part of the contractor and reflects the contractor's keenness to win the contract by keeping overheads and profits tight.

A typical bill of quantities for a building contract is divided into the following sections.

- preliminaries (overheads such as insurance, plant, tools, etc.)
- excavation and earthwork
- concrete work
- brickwork and blockwork
- drainage
- roofing
- carpentry and joinery
- plastering
- plumbing
- electrical
- painting and decorating
- prime cost and provisional sums (see contract clauses under the subheading 'contract conditions')

### Specification

This is a precise description of all the essential information and job requirements that will affect the price of the proposed work which cannot be

13

CONWAY TRUNKING LTD
Unit 8
Gladstone Park
Lancs LA4 TT4
Tele 0883–565543, Fax 0883–0998990

Quotation No. 9930894

Client: Fort William Steeple
Hanger Lane
Bucks

Date: 19-12-94
Project: Chubb House
(Ground Floor)

| Ref No. | Colour | Quantity | Description | Nett Trade Price | Per | Total |
|---|---|---|---|---|---|---|
| ETB1 | WHI | 33m | 13 Lengths Sterling Base 3-Compartment trunking | 11.35 | metre | 374.55 |
| ETL1 | '' | '' | Lid | 2.46 | '' | 81.18 |
| ETAC | '' | 66 | Angle Covers | 2.46 | '' | 162.36 |
| ECP1 | '' | 4 | Couplers | 3.47 | each | 13.88 |
| EECP1 | '' | 6 | End caps | 2.33 | '' | 13.98 |
| EIBP1 | '' | 6 | Internal bends | 4.60 | '' | 27.60 |
| EXBP1 | '' | 2 | External bends | 4.60 | '' | 9.20 |
| EFT1 | '' | 1 | Flat tee | 58.45 | '' | 58.45 |
| ECP1 | '' | 2 | (Couplers for above) | 3.47 | '' | 6.94 |
| ESSB1 | '' | 20 | Single Boxes (Ref.BT Comp) | 1.63 | '' | 32.60 |
| ESSB2 | '' | 16 | Twin Boxes (Ref twin SW/SL) | 2.29 | '' | 36.64 |
| FM15 | '' | 8 | Standard mounting plate | 1.67 | '' | 13.36 |
| RJ45 | '' | 8 | BT plate | 3.30 | '' | 26.40 |
| FM02 | '' | 8 | Half blanks | 1.17 | '' | 9.36 |

Final total (Nett trade)                                                                                           £866.50

Signed  *J. Smith*

Title  *Designer Consultant*                                         Date *20/12/94*

*This quotation is given in good faith and we would recommend that the requirements stated should be checked prior to placing an offical order. We cannot accept responsibility for any errors which may have been incurred in formulating this quotation.*

Figure 1.12   Bill of quantities (VAT-free prices)

shown on drawings. Building specifications are prepared by an architect and include a description of the site and restrictions, availability of services, description of materials i.e. quality, size, finish, etc. It will also include a description of workmanship and requirements such as site clearance, making good on completion, nominated suppliers and sub-contractors.

An electrical specification is produced by an electrical consultant. He has to ensure that it correctly identifies the work that the electrical contractor is to undertake. For it to be meaningful, the standards and regulations (statutory and non-statutory) that relate to the proposed installation must be clearly stated. For example:

The contractor shall ensure compliance with the *Health and Safety at Work etc Act 1974* and shall give the client prior to starting the work, a copy of his safety policy statement and details of his arrangements for securing compliance with the terms of the statement.

Compliance with the safety rules and guidance in the latest edition (16th Edition 1991) of the *IEE Wiring Regulations* shall be assured throughout the contract.

The specification generally consists of two sections, one being the requirements for standards of workmanship and the other being particular requirements for the installation. In the standard section the electrical contractor will be required to use materials for the installation that are suitable for their intended purpose. For example:

All cables shall be to a relevant British Standard and be of 300/500 V rating; be colour coded to the latest requirements of the *IEE Wiring Regulations* and where buried in walls, floors, ceilings, etc. shall be protected by galvanised capping or high grade pvc capping fastened to the walls.

Where steel conduit is to be used, it should be of the appropriate size, and manufactured in accordance with relevant British Standards (e.g. BS 31, BS 6063, BS 6099 etc.) using the correct tools (e.g. bending machine, stocks and dies, reamer tool, etc.)

The electrical contractor will be required to ensure that the number of bends between draw-in boxes comply with the *IEE Wiring Regulations* and that the appropriate number of boxes and fixings are used (see Section 5 and Appendix 1 of the *IEE Guidance Notes No. 1, Selection and Erection of Equipment*).

In the particular section of specification the contractor will be advised, for example, that a conduit of galvanised finish of 25 mm diameter is to be installed between points A and B for the purpose of carrying X number of cables for a specific service.

The consultant has the responsibility of ensuring that his specification correctly identifies and details the works the electrical contractor is to undertake. The electrical contractor, in turn, has the responsibility to ensure that the works are carried out strictly in accordance with the specification. If this is not possible, then a variation will have to be issued to the electrical contractor to carry out the works in a revised manner. An explanation of the word 'variation' is given below under contract clauses.

*Schedules*

In main building work, these are used to record repetitive design information about a range of similar components, e.g. doors, windows, radiators, etc. In electrical contracting terms this might be concerned with distribution boards, luminaires, equipment manufacturers and even drawing symbols. This information is essential when preparing estimates and tenders and is also useful when measuring quantities, locating work and checking deliveries of materials and equipment.

*Conditions of contract*

A contract is an agreement between parties which contemplates and creates an obligation. In order to create an obligation the agreement must be legal and the parties involved must have the capacity to enter into a contract. The contract should cover all aspects of terms and conditions and provide protection against sharp practice. It should be an assurance that the work and payment will be carried out in an honest and professional manner.

The first stage in forming a contract is when an offer is made by the contractor for the customer to accept, should he wish to do so. The contractor's offer is seen as an undertaking to carry out the work in return for a promise by the customer to pay the price for his work. It is the acceptance of the offer that completes the contract, but until it has been accepted there is nothing to bind the parties. Both, the offer and acceptance may be in writing or spoken but the latter type of contract is only satisfactory if disputes do not arise. If they do, there is no written evidence to resolve them.

**Note:** Certain conditions may prevent a contract from being made or accepted, such as withdrawal of the person(s) making the offer; lapse of offer time or the customer rejecting it.

The standard form of contract used will depend on the type of customer or client (local authority, public limited company or private individual). The form of contract will also depend on the size and type of work to be undertaken and the contract documents themselves. The building industry may use the Building Employers Confederation (BEC) form of contract or the Joint Contractors Tribunal

(JCT) standard forms of building contract (JCT 80). This latter form is used considerably in electrical contracting work and has already received fourteen amendments. Ammendment 10 concerned procedures for the nomination of a subcontractor by the architect. This existed because some architects wished to exercise control over the selection of specialist contractors who became subcontractors to the main building contractor (see Figure 1.13). These procedures were intended to reduce the incidence of disputes on nominated subcontracts. Amendment 10 created a new Standard Form of Tender (NSC/T) which reduced the need to pass earlier types of tender form between parties.

Two other types of contract used by electrical contractors are (i) the model forms for domestic subcontractors, DOM/1 and 2, and (ii) the standard form for named subcontractors, NAM/SC, used for the JCT Intermediate Form of Building Contract (IFC 84).

It is often the case with large contracts that the customer states the terms that are to be offered and the contractor has to tender on the basis of these requirements. These requirements are set out in various clauses and subclauses, such as:

the tender remains open for 30 days unless previously withdrawn;
acceptance of the tender includes acceptance of the terms and condition;
the works are to carried out in a workmanlike manner;
good quality materials and equipment must be used;
qualified labour, paid the correct rates, is to be used;
adequate supervision is to be provided;
the contractor must work to a programme;
offer a guarantee for his work and be paid on submission of his interim certificates;
additional monies will be payable for increased costs should the contract be on a fluctuating basis.

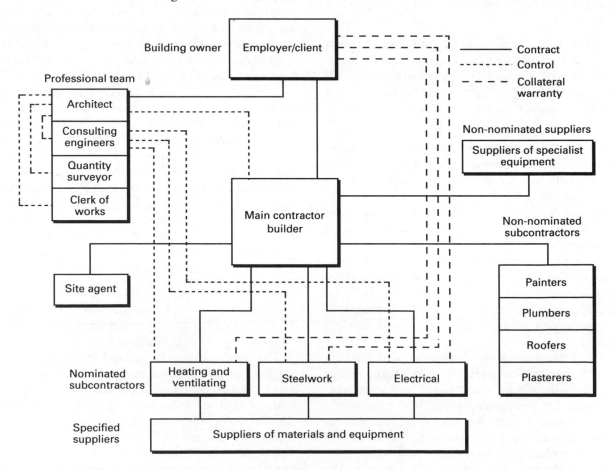

Figure 1.13 Nominated and non-nominated subcontractors

Some important terms found in contract clauses are listed below but you can find more about these in chapter 5 of the author's *Electrical Installation Technology 3*.

**Programme.** The tender is submitted on the understanding that subject to the receipt of information, the proposed work will be completed by a certain date and that a detailed programme of the works will be drawn up and be subject to agreement between the subcontractor and the main contractor.

**Variations.** A client sometimes changes the requirements on some aspect of work being undertaken or an incorrect detail is discovered on a drawing or in the specification. Any modification of the work requires a variation order to be made. This is generally advised by the consulting engineer to the architect who then issues an architect's instruction (AI) to the main contractor to carry out the extra work in accordance with the terms and conditions of the contract. Figure 1.14 shows a typical form for advising variants. Prior to starting

---

**ADVICE OF VARIATIONS**

To ..............................      From ..............................

Contract ..........................      Job No. ..........................

Date ..............................

Today the following verbal instruction as a variation to the contact was given.

Would you therefore please confirm it in writing.

To ..............................

**VERBAL INSTRUCTIONS**

Add .............................................................

...................................................................

...................................................................

...................................................................

...................................................................

...................................................................

Omit ...........................................................

...................................................................

...................................................................

...................................................................

...................................................................

...................................................................

Instruction issued by .................      Position ..........................

Instruction issued to .................      Position ..........................

Signed ..........................

Figure 1.14    Advice of variations

work on site the electrical contractor has usually agreed a schedule of rates with the building quantity surveyor.

**Schedule of rates.** These are unit costs appearing in a bill of quantities. They are normally required for abstract items and are produced to give a total cost. In most instances, such rates will become the basis for variations and in consequence will not be based on the gross value of the estimated costs because of, for example, higher overhead costs in buying/invoicing smaller quantities in which the same charges are made for small deliveries as for bulk deliveries.

**Interim payment.** A periodic payment (14 days or monthly) made to the contractor by the client based on the work done and material purchased by the contractor.

**Retention.** A percentage sum of money (5%) that is retained by a client until the end of an agreed defect liability period (i.e. a period of normally six months). The contractor will be entitled to the retention payment after any defects have been rectified to the architect's satisfaction.

**Prime cost sum.** This is an amount of money to be included in the tender for work, services or materials provided by a nominated subcontractor, supplier or statutory body.

**Provisional sum.** A sum of money to be included in the tender for work which has not been defined to cover the cost for any unforeseen work.

*Contract of employment*

Under the *Employment Protection (Consolidation) Act, 1978*, as amended by the *Employment Acts, 1980 and 1982* employers must give all their employees, who are employed for 16 hours or more per week, a written statement containing the conditions of their employment. This has to be within 13 weeks of starting work. The statement must contain an additional note on disciplinary and grievance procedures, specifying any disciplinary rules that apply. Employees who have been continuously employed for one month or more (for the weekly hours mentioned) are given statutory rights to the length of notice upon termination of employment and minimum pay during notice. Either party may waive his right to notice or to mutually accept an agreed payment in lieu. The Act does not prohibit an employer instantly dismissing an employee at any time for serious misconduct, misdeed or bad workmanship.

**Note:** Most new entrants to the electrical contracting industry come in as junior electrical apprentices and as such enter into a training contract with their employer. Employers who are members of the JIB have to abide by the JIBs National Working Rules and Industrial Determinations. For reference, see *Code of Good Practice – Employment of Operatives* in the *JIB Handbook*.

## Electrical organisations

There are a number of organisations directly involved in the electrical engineering and electrical contracting industry, namely:

● The Electrical Contractors Association (ECA)
● Joint Industry Board for the Electrical Contracting Industry (JIB)
● Amalgamated Engineering and Electrical Union (AEEU) (which incorporates the Electrical, Electronic, Telecommunication and Plumbing Union (EETPU))
● National Inspection Council for Electrical Installation Contracting (NICEIC)
● JT Limited (JTL)
● Electricity Training Association (ETA)

### *Training and qualifications*

Figure 1.15 shows various organisations involved in funding, administration and training. All funding is through the Training and Enterprise Education Directorate (TEED) based in the Department of Employment.

NVQs are competence-based awards (**National Vocational Qualifications**) which are associated with a particular occupation. They are offered at various levels and signify that a person is competent to perform a specified range of work-related tasks.

In electrical installation engineering there are currently three levels which provide NVQs, namely Levels I, II and III:

*Level I – Fabricating and Fixing Electrical Cable Supports/Preparing and Fixing Cables* which consists of six units each and require a candidate to prove competence in a range of work activities most of which are routine and predictable.

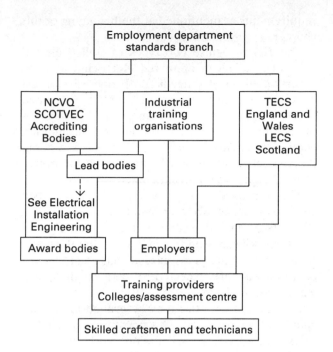

**Figure 1.15    Structure of industrial training**

*Level II – Installing Electrical Systems and Equipment* which consists of six units each requiring a candidate to prove competence in a significant range of varied work activities, some of which are complex and non-routine and where there may be some individual responsibility.

*Level III – Installing and Commissioning Electrical Systems and Equipment* consists of eleven units in which the work activities are complex and non-routine. Guidance of other persons may be required. Success at this stage is the minimum requirement to obtain skilled electrician status.

The **City and Guilds of London Institute** was formed in 1878 and is incorporated by Royal Charter. It has a long record of administrating craft courses and works closely with representatives from industry, commerce and training organisations in order to develop a wide range of assessment and qualifications. All NVQ candidates will register with City & Guilds through an approved centre.

The **Construction Industry Training Board** for Northern Ireland was set up as a Statutory Board in 1964 in order to providing an adequate number of well trained people in industry.

*Electrical Contractors Association*

The ECA was founded in 1901 to develop the electrical installation engineering trade. It was not until the *Trade Union Act, 1913* that it was able to negotiate conditions of employment and salaries in the electrical contracting industry. The Association's aim is to ensure that electrical installation work is undertaken by qualified people to high standards of quality and safety and to terms that are fair to the client and installer.

All ECA members' firms (over 2,300) are members of the JIB and must comply with the JIBs *National Working Rules*. Electrical installations carried out by ECA firms carry an independent guarantee against bad workmanship. To be a prospective member, an electrical contractor must have been operating successfully for at least three years and the standard of his work must be to the current edition of the national *IEE Wiring Regulations* and other relevant standards.

The Association has members serving on the **Electrical Installation Engineering Industry Training Organisation (EIEITO)** which is the lead body responsible for writing standards of performance for National Vocational Qualifications (see Figure 1.16).

The EIEITOs primary aim is to ensure that there are sufficient skilled people in the industry to meet the industry's current and future needs. The EIEITOs joint awarding body consists of four awarding body partners, namely, the Joint Industry Board, City & Guilds of London Institute, the Construction Industry Training Board for Northern Ireland and the Electricity Training Association. This group is responsible for producing 'field evidence records' which form part of a portfolio of evidence for people undergoing training.

*Joint Industry Board for the Electrical Contracting Industry*

The JIB was established in 1968 by the ECA and EETPU. Membership of its National Board consists of equal representation from the AEEU and ECA, under an independent chairman.

The JIBs main function is to regulate relations between employers and employees within the electrical contracting industry. It does this through its National Working Rules and Industrial Determinations. Some of the services provided to its members are: grading of operatives, training of apprentices and industrial relations, productivity,

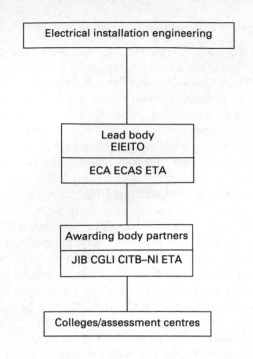

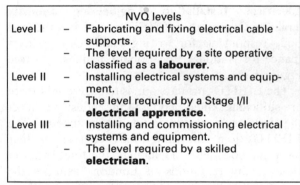

| NVQ levels | | |
| --- | --- | --- |
| Level I | – | Fabricating and fixing electrical cable supports. |
| | – | The level required by a site operative classified as a **labourer**. |
| Level II | – | Installing electrical systems and equipment. |
| | – | The level required by a Stage I/II **electrical apprentice**. |
| Level III | – | Installing and commissioning electrical systems and equipment. |
| | – | The level required by a skilled **electrician**. |

Figure 1.16 NVQ awards in electrical installation engineering industry

health, safety and welfare. A brief summary follows, but more about these topics and other services can be found in the JIBs Handbook.

**Grading of operatives.** A standard of competence is accepted throughout the industry. Installation operatives are individually graded on technical qualifications and practical experience. At senior level is a **technician**. This person must have at least five years experience as an approved electrician and be at least 27 years of age. He/she must be in possession of the City & Guilds electrical installation work Course 'C' certificate and have the knowledge and ability to run electrical installations economically to achieve a high level of productivity. This person must also have a good understanding of relevant statutory and regulatory

requirements as mentioned at the beginning of this chapter.

The JIB recognise two types of skilled electrician, one called an **approved electrician** and the other called an **electrician**. Both operatives must have been registered as apprentice, obtained practical Achievement Measurement 2 and completed the City and Guilds electrical installation competences Part II certificate. If this is not possible, the person has to satisfy the grading committee of their experience and suitability.

The approved electrician must have had two years experience working as an electrician or be 22 years of age, whichever is the sooner. His/her practical skills and understanding of statutory and regulatory requirements exceed that of an electrician and they are expected to work on their own without supervision.

A stage 2 apprentice may be graded an electrician at 21 years of age. This person is expected to carry out electrical installation work efficiently and have a knowledge of the JIB rules and safety regulations.

Under its 'cable agreement rules', the JIB recognises the employment of **labourers** to assist in the installation of cables and do other unskilled work under supervision. For example, they may be employed for pulling in heavy cables or used with skilled operatives to erect tray work, fix brackets, clip cables, etc. They must not be used to reintroduce pair working.

**Training of apprentices.** This is through the JIBs '1983 JIB Training Scheme'. A young person meeting the Board's entry and selection requirements is classified as a **junior apprentice**. He/she will undergo training and education that currently incorporates 12 weeks off-the-job practical training and 12 weeks college-based technical education. After a 12-month period and satisfactory progress on site and in college the person's job description changes to that of **senior apprentice**.

It takes two further years (stages 1 and 2) to complete the apprenticeship programme and successfully emerge as a skilled graded **electrician** (see Figure 1.17).

**Industrial relations.** This is basically concerned with work co-operation between employers and employees. Without a good working relationship between both parties, conflict inevitably occurs. Disputes are usually outcomes of grievances that have been ignored and occur when workers feel that an

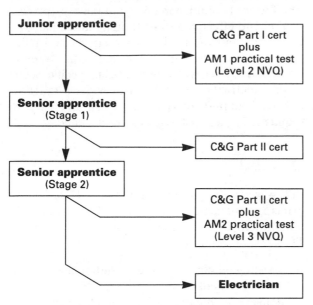

**Electrical apprenticeship route to become a JIB graded electrician**

Training stages                         Awarding stages

Junior apprentice → C&G Part I cert plus AM1 practical test (Level 2 NVQ)

Senior apprentice (Stage 1) → C&G Part II cert

Senior apprentice (Stage 2) → C&G Part II cert plus AM2 practical test (Level 3 NVQ)

→ Electrician

Figure 1.17    JIB training structure for electrical apprentices

employer has taken a decision without adequate consultation or agreement. In the electrical contracting industry, industrial relations matters are progressed by the JIB. Dispute procedures are fully dealt with in the their National Working Rules (see Note below).

It is important that these rules are observed and attempts made to resolve a dispute locally.

If the dispute involves an employee he/she should, firstly, report to their immediate supervisor. If a satisfactory solution is not found, then the matter should be taken up with the employee's job/shop representative who will approach the supervisor concerned. If the dispute is still unresolved, each party should refer the matter to their representative organisations. Failure at this stage means that it will have to be dealt with by the local Regional JIB (RJIB) disputes committee in the form of a full written report.

Although most disputes and grievances are resolved at this stage, an appeal may be lodged against the RJIB's decision. This has to be in writing to the secretary of the JIB within twenty-eight days. It is then dealt with and finally resolved by a committee of the JIBs National Board. On other matters such as unfair dismissal and discipline, you should consult the JIB handbook.

**Note:** The JIB disputes procedure is exempt from the Unfair Dismissals Provisions of the *Employment Protection Act*. Non-JIB operatives would have to seek the services of the *Advisory Conciliation and Arbitration Service (ACAS)*. This organisation is concerned with industrial disputes and not contractual disputes.

**Productivity.** This is concerned with the provision of services provided by the JIB to the electrical contracting industry. It covers such things as a national library of standard data on work studied times, training, plant, tools and equipment etc.

**Welfare.** Welfare covers a number of benefits to operatives such as holidays, sick pay, life and disability assurance, membership, medical health care and benefits during unemployment.

**Health, safety and welfare.** These are covered in a Code of Good Practice issued to new entrants.

### Amalgamated Engineering and Electrical Union

The AEEU was formed in 1992 and is a relatively new trade union consisting of the Amalgamated Engineering Union (AEU) and Electrical, Electronic, Telecommunications and Plumbing Union (EETPU) which is mentioned below. At present, both these long established unions operate as independent sections with eleven representatives from each serving on its Executive Council.

Some of the Union's objects are: to promote the general economic and social well being of its members; to represent members' interests and improve their working conditions; and to provide financial benefits and legal assistance to its members. The Union's services and benefits can be found in their booklet called *At Your Service*.

### Electrical, Electronic, Telecommunication and Plumbing Union

The EETPU was established in 1968 and is an amalgam of two craft unions, notably the Electrical Trade Union (ETU) and Plumbing Trade Union (PTU). The EETPU's principal objects are to organise all manual, operative, technical, administrative, supervisory, professional and managerial workers in the industries specified by its title including mechanical engineering services and allied trades, and professions in any other industry. The objects of the Union extend to that of improving wages and conditions of its members, providing legal support, provision of educational facilities, convalescent, sickness and financial hardship. It is now operating as a separate section within the AEEU. Figure 1.18 shows its industrial structure

### National Inspection Council For Electrical Installation Contracting

The NICEIC was formed in 1956 to protect consumers against unsafe and unsound electrical installations. The Council operates as an independent voluntary, non-profit-making organisation for electrical installation contractors. As a regulatory body it is supported by ETA and other organisations. Electrical contractors accepted on its Roll are approved for whatever electrical installation work they choose to carry out. Once a contractor is enrolled with the Council they will receive a regular annual visit by an inspection engineer. This is to see if the contractor continues to comply with the Council's rules and that the required standard of installation work is being maintained.

**Note:** The standard of work required by an electrical contractor must conform to the *16th Edition of the IEE Wiring Regulations* as well as codes of practice published by the British Standards Institution.

### Electricity Training Association

ETA was established in 1990 and is the lead body and training organisation for the **Electricity Supply**

**Industry** (generation, transmission, distribution and utilisation). Before the *Electricity Act of 1989*, ETA was called the Electricity Council. The aims of ETA are twofold, namely: (i) to raise the profile, quality and effectiveness of training in the electricity sector, and (ii) to develop a strong and independent presence within the industrial training organisation that will represent the specific needs of the electricity sector. ETA provides a range of optional support services of which the City & Guilds 2320 and 2360 courses are included.

### JT Limited

JTL was set up by the ECA and AEEU and is the principal managing agent for the training of apprentices in the electrical contracting industry. JTLs responsibilities include:

- assisting employers in the recruitment, selection and placement of apprentices and technicians;
- delivery of quality training at all levels under the JIB 1983 Apprenticeship Scheme;
- the JIB 1989 Adult Craft Training Scheme.

These responsibilities extend to monitoring and assessing the progress of apprentices and also organising practical achievement measurement tests. For further information about JTL, see their booklet called 'Training for Industry' which provides useful guidance notes to employers, electrical apprentices and training officers.

## Industrial agreements

These are made at a national or local level to cover aspects of employment such as scope of work, wages, type of work, method of working, travelling allowance, etc. The JIB Handbook covers national agreements made for installations of cables in buildings, shipwork, onshore work and specified engineering construction sites.

Where a construction site has its own special requirements a local agreement can be formed. This might be to determine the grades of labour used or to ensure that local labour is used for unskilled work. It might also be formed if the location of a site is in a residential area and noise output is restricted to certain agreed times.

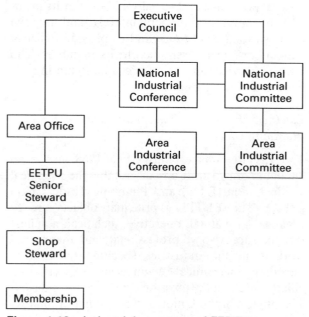

Figure 1.18   Industrial structure of EEPTU

## Elements of electrical contracting management

In order to meet the needs of clients and customers an electrical contractor has to conduct his business in a logical and efficient manner taking into consideration:

- development strategies;
- design, control and deployment of resources;
- testing, commissioning and maintenance;
- contributions to the competitiveness and effectiveness of his business.

Development strategies are concerned with identifying factors that affect the promotion and improvement of services and resources. Design is concerned with assessing installation requirements and evaluating various constraints both technical and physical. It will involve the preparation of forms of contracts, drawings, specifications, bills and schedules that have already been mentioned and it should also involve information and instruction to clients and customers in the form of installation manuals (see Chapter 3).

The procurement, control and deployment of resources has to consider not only human, capital and financial resources but also material and equipment resources. In the latter case, the contractor will need to establish both current and future requirements of material and equipment, select suppliers, contract for the supply of these resources, process them and deploy them to their respective points of utilisation.

### *Site organisation*

Good site organisation usually reflects improved productivity, better industrial relations and an improved profit margin. A well run building site usually inspires trade subcontractors to operate efficiently, whereas a poorly run site results in delays, slow progress, indecision and is often vulnerable to accidents. Such factors are demoralising to the workforce, particularly junior operatives learning essential skills. It is extremely important for the main building contractor and service subcontractors to appoint well qualified and experienced site representatives to act for them.

As for an electrical subcontractor, the person in charge of running a medium-to-large site is usually a foreman electrician (site supervisor). On very large sites the foreman may delegate sections of work to chargehand electricians.

The foreman's responsibility to his installation team is to show leadership, authority and fairness. It is essential for him to have a first class labour/management relationship with his team, making sure that operatives know the tasks that have to be performed. With operative skills varying considerably, the foreman will delegate work schedules to those who have the experience to complete it in the allotted time. It gives considerable satisfaction to operatives if they are able to see the completion of their work on one site.

Figure 1.19 shows the tasks to be performed by the design team and installation team. If the electrical installation has been properly planned the first requirement is to establish facilities on site.

### *Site office*

This is often a site hut and should be of adequate size for the number of operatives employed. It should be of safe construction with a door that can be locked and it should be given a temporary supply of electricity and a telephone point. It should also contain welfare facilities for operatives to wash their hands, go to toilet and make tea. Sometimes these facilities are provided by the main building contractor or they may be allocated in some part of the building premises, depending on its stage of construction. Ideally, the site hut

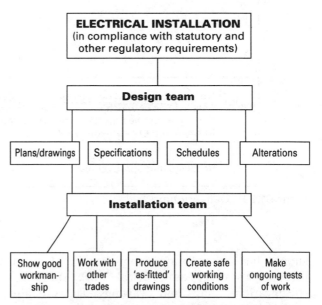

Figure 1.19   Design and installation teams for an electrical installation

23

| J. LUMINAIRES (U.K.) LTD |
|---|
| Electrical Wholesalers |
| Unit 9, Elm Storage Depot, Kessington, Beds |
| Tele Beds 113300 |

**CREDIT SALE DELIVERED**

| Customer | Date | Order No |
|---|---|---|

**Site Address**

| Haulier | Driver | Vehicle No. |
|---|---|---|

| Quantity | Description | Code |
|---|---|---|
| | | |
| | | |
| | | |
| | | |
| | | |
| | | |
| | | |
| | | |

Comments

Goods are supplied on the basis of this Company's Conditions of
Sale, copies of which are available on request. Notification of
damage must be reported in two days and errors must be notified
to the Company immediately

Received and approved by .................................................................

Figure 1.20   Delivery note

should have a separate room for the electrical
foreman to carry out clerical work, study the
specifications, plans and drawings. This will also
allow him to discuss matters with his installation
team and any visitors to the site.

The foreman needs to keep a **site diary** to record
site meetings, important telephone messages and
telephone numbers. The diary can also be used for
keeping records of damaged material, important
dates, site attendance and accident information.

## Delivery of materials and equipment

Regarding materials and equipment on site, it should be the aim to hold a minimum stock in hand in the stores to ensure that work is never held up. In practice, the electrical firm's main office usually sends written orders to the suppliers, consisting of a description of type, catalogue number, quantity and delivery location. Materials delivered directly to the site are accompanied by a delivery note that needs signing by the person receiving the goods (see Figure 1.20).

The materials listed should be checked against the original order form and a note made of any items that are incomplete. As a safeguard, both, site operative and delivery driver should sign the delivery note indicating the items that are missing. If the supplier is out of stock, be aware that the items missing may already be typed on the delivery note. Damaged items should be sent back to the supplier and the supplier notified, preferably within a few days. Once the goods have been delivered to the site, the supplier will forward an invoice to the firm's office, requesting payment.

## Programme of work

Whether a project is large or small the main objective is to bring the work involved to a successful conclusion. To achieve this final outcome, there is a need for everybody working on site to appreciate and respect the requirements of others. This calls for close liaison with those individuals charged with organising the work in their particular trade. It is often the case with large projects that the main building contractor prepares a programme showing the sequence of work operations of various trades.

The method most frequently used is called a **bar chart** (see Figure 1.21). It provides immediate visual information of the various work operations needed to complete the project. It also indicates when it is suitable to order materials and equipment and it can be used to show the number of operatives needed for different work operations.

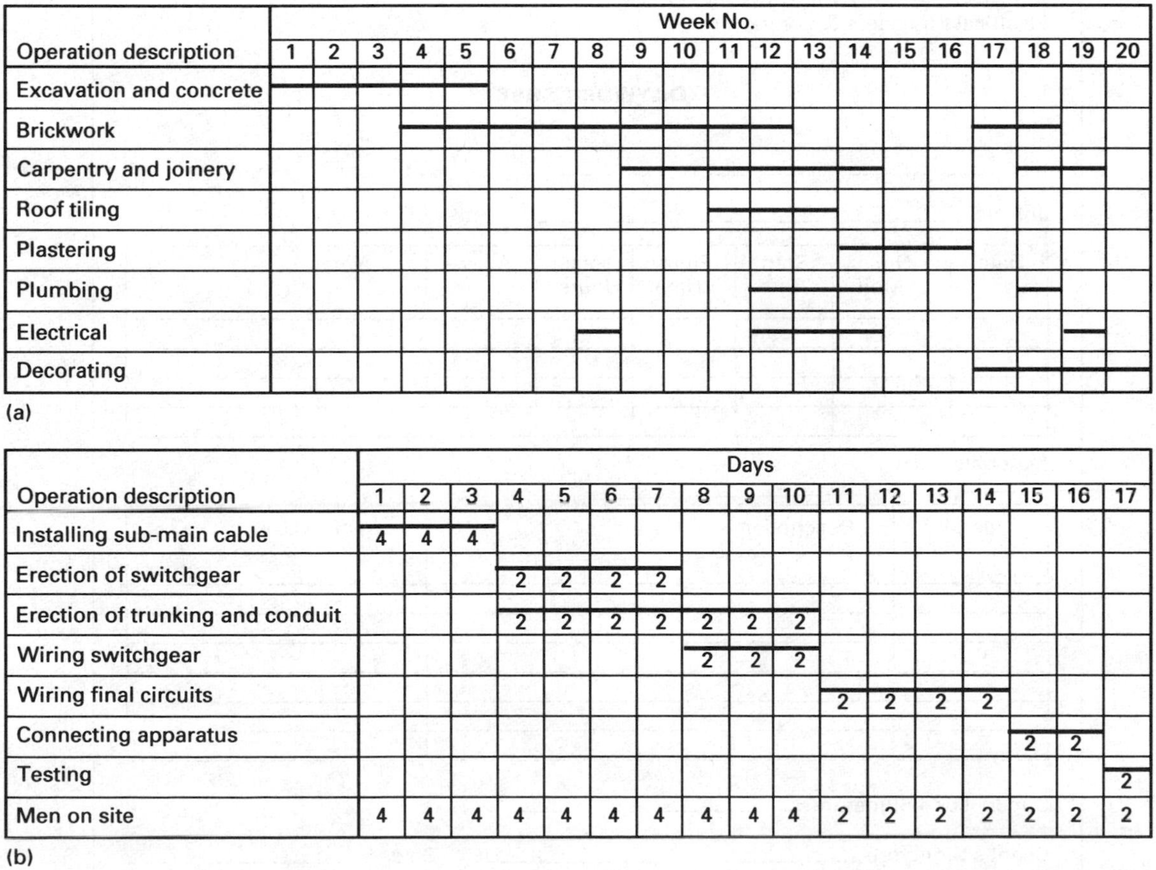

| Operation description | Week No. | | | | | | | | | | | | | | | | | | | |
|---|---|---|---|---|---|---|---|---|---|---|---|---|---|---|---|---|---|---|---|---|
| | 1 | 2 | 3 | 4 | 5 | 6 | 7 | 8 | 9 | 10 | 11 | 12 | 13 | 14 | 15 | 16 | 17 | 18 | 19 | 20 |
| Excavation and concrete | ▬ | ▬ | ▬ | ▬ | ▬ | | | | | | | | | | | | | | | |
| Brickwork | | | | | ▬ | ▬ | ▬ | ▬ | ▬ | | | | | | | | ▬ | ▬ | | |
| Carpentry and joinery | | | | | | | | ▬ | ▬ | ▬ | ▬ | ▬ | | | | | | ▬ | ▬ | |
| Roof tiling | | | | | | | | | | | ▬ | ▬ | | | | | | | | |
| Plastering | | | | | | | | | | | | | ▬ | ▬ | ▬ | | | | | |
| Plumbing | | | | | | | | | | | | | | | | ▬ | | | | |
| Electrical | | | | | | | ▬ | | | | ▬ | ▬ | ▬ | | | | | ▬ | | |
| Decorating | | | | | | | | | | | | | | | | | ▬ | ▬ | ▬ | |

(a)

| Operation description | Days | | | | | | | | | | | | | | | | |
|---|---|---|---|---|---|---|---|---|---|---|---|---|---|---|---|---|---|
| | 1 | 2 | 3 | 4 | 5 | 6 | 7 | 8 | 9 | 10 | 11 | 12 | 13 | 14 | 15 | 16 | 17 |
| Installing sub-main cable | 4 | 4 | 4 | | | | | | | | | | | | | | |
| Erection of switchgear | | | | 2 | 2 | 2 | 2 | | | | | | | | | | |
| Erection of trunking and conduit | | | | 2 | 2 | 2 | 2 | 2 | 2 | 2 | | | | | | | |
| Wiring switchgear | | | | | | | | 2 | 2 | 2 | | | | | | | |
| Wiring final circuits | | | | | | | | | | | 2 | 2 | 2 | 2 | | | |
| Connecting apparatus | | | | | | | | | | | | | | | 2 | 2 | |
| Testing | | | | | | | | | | | | | | | | | 2 |
| Men on site | 4 | 4 | 4 | 4 | 4 | 4 | 4 | 4 | 4 | 4 | 2 | 2 | 2 | 2 | 2 | 2 | 2 |

(b)

Figure 1.21 Bar charts: (a) simple building trade operation; (b) electrical installation showing work activities of persons on site

The bar chart is constructed with individual tasks listed vertically and in the order the work will be carried out. The time taken for each task to be complete is marked as a single horizontal line, expressed in months, weeks or days.

The electrical foreman may receive a bar chart from the main builder indicating the dates when various sections of his work can commence. For example, in Figure 1.22(a), week 8 is the time to run the main cable to the intake position and weeks 12–14 are the times to carry out first and second stage fixings. The first stage is when the walls of the building are up to flat roof level and shuttering begins. This allows the electricians to install galvanised metal conduit to lighting points and the switch drop positions. The second stage fix is when the plastering work is completed and the conduit can be wired and accessories terminated. A third stage is shown (week 19) for testing, commissioning and handing over the installation.

## Site records

All electrical contracting work involves documentation to be completed. The electrical foreman will have to sign daywork sheets and time sheets and write frequent reports.

**Daywork** is the unavoidable work outside the scope of the contract, used to cover extra work that has to be paid for (see variations). It is important that the information regarding materials is accurate and the names of all site operatives and their hours worked are included on the sheet. The foreman will have to obtain a signature of approval by the client's representative and then return it to his main office for pricing up and submission to the client/main contractor for payment. Figure 1.22 is a typical daywork sheet.

**Time sheets** provides the employer with a permanent record of labour on site. They are normally a

---

ALPHA-BETA ELECTRIC COMPANY (U.K.)
Electrical Engineers & Contractors

### DAYWORK SHEET

Client _____

Job No. _____

| Date | No. of Men | Start Time | Finish Time | Total Hours | Allow | Notes |
|------|------------|------------|-------------|-------------|-------|-------|
|      |            |            |             |             |       |       |
|      |            |            |             |             |       |       |
|      |            |            |             |             |       |       |
|      |            |            |             |             |       |       |

Materials

| Quantity | Description | Office Use |
|----------|-------------|------------|
|          |             |            |
|          |             |            |
|          |             |            |
|          |             |            |
|          |             |            |
|          |             |            |

Supervisor's Signature _____

Client's Signature _____ Date _____

Figure 1.22   Daywork sheet

```
ALPHA-BETA ELECTRIC COMPANY (U.K.)
Electrical Engineers & Contractors


                        TIME SHEET


Name _____

Week Ending _____
```

| Day | Job No. | Start Time | Finish Time | Total Hours | Travelling Time | Fares Milage |
|-----|---------|------------|-------------|-------------|-----------------|--------------|
| Sun |  |  |  |  |  |  |
| Mon |  |  |  |  |  |  |
| Tues |  |  |  |  |  |  |
| Wed |  |  |  |  |  |  |
| Thurs |  |  |  |  |  |  |
| Fri |  |  |  |  |  |  |
| Sat |  |  |  |  |  |  |
| Totals |  |  |  |  |  |  |

```
Employee's Signature: _____

Supervisor's Signature: _____   Date:   _____
```

Figure 1.23   Weekly time sheet

weekly document signed by each site operative and will include the number of hours worked together with any travelling time, fares, etc. Figure 1.23 is a typical time sheet.

**Reports** are written not only to show progress of work on site but also to show if early action is needed to deal with any problems that might arise. This could be due to slow progress being made on site as a result of delay in materials arriving or work that cannot start and labour becomes surplus to requirements. Reports, like minutes of meetings, are vital documents to keep for reference purposes and enable claims to be substantiated.

# EXERCISE 1.2

1. (a) With reference to Figure 1.9, state the length and breadth of the room using the 1:50 scale.

   (b) Using typical manufacturers catalogues, make a list of all the equipment shown on the drawing and where possible, quote British Standard numbers.

   (c) Assume the wiring systems are steel trunking and steel conduit (on surface), estimate the quantity required. Allow 2 m drops to lighting switches, motor starters and 3.5 m to all socket outlets.

2. Figure 1.24 shows the supply to a domestic consumer's premises provided with a TT earthing system. Draw the following diagrams.

   (i) a line/block diagram showing the sequence of control into the premises using, where appropriate, BS 3939 installation graphical location symbols.

   (ii) a circuit diagram of the complete system including the earthing arrangements. Label all circuit conductors and state the sizes that should be used for a 100 A supply.

27

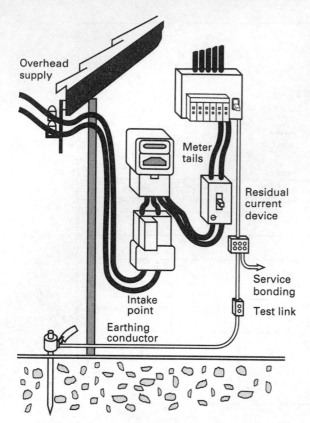

Figure 1.24    TT earthing system

**3.** (a) Explain what is meant by the term 'bar chart' and state its advantages in a large project.

   (b) Produce a bar chart from the following work activities for installing floodlighting. Each activity is expressed in days (in brackets).

   1  design and plans (2)
   2  await cables (2)
   3  await floodlighting control gear (1)
   4  await floodlights (3)
   5  await columns (7)
   6  await column control gear (1)
   7  lay column cables (3)
   8  install floodlight cables (1)
   9  erect floodlights (1)
   10  install floodlight control gear (1)
   11  connect floodlights (1)
   12  erect columns (2)
   13  install column control gear (1)
   14  connect columns (1)
   15  test and commission (1)

**4.** Briefly explain what constitutes a contract between two parties and describe the main advantage to an electrical contractor of using a standard form of contract rather than a contract devised by a client.

**5.** With reference to the installation shown in Figure 1.9, it is found that the BS 3036 protective devices in the fuseboard are the wrong type and need to be replaced by BS 1361 cartridge fuses. On the assumption that the fuseboard is already connected complete the contract instruction form Figure 1.25(b) to carry out the alteration. Estimate the cost of carrying out the work. A sample form is shown in Figure 1.25(a).

**6.** With regard to electrical contracting work on a construction site, prepare notes on the following topics.
   (i)   accommodation for operatives
   (ii)  access equipment
   (iii) first aid
   (iv)  site meetings

**7.** As a representative of your employer in a work situation, write notes on good customer relations and site behaviour covering small and large projects.

**8.** Explain the procedure for completing the following site documents.
   (i)  time sheet
   (ii) daywork sheet

**9.** Briefly state the role of the following personnel.
   (i)   architect
   (ii)  clerk of works
   (iii) quantity surveyor
   (iv)  contracts manager
   (v)   structural engineer

**10.** A number of electrical accidents are caused by failure to isolate the supply. State the statutory and regulatory requirements concerning isolation and describe the procedure for carrying it out.

# Contract instruction

XYZ Group
Elm House,
Wix Corn Rd,
Paverly
Hants.

Alterations to
Engineering services

**Site address**
2 Hill Lane
Boxted Park
Glam.

Description of work

**Contractor address**
W.F. Tilley
108 Barnes Rd
Boxted, Cambs

| | |
|---|---|
| Job reference | 5604/EP11262/AB/OLM |
| Issue date | 14 December 1994 |
| Instruction no | SEVEN |

The contract sum will be adjusted where applicable in accordance with the terms of the relevent contract condition.

Instruction

7.01 Please instruct Alpha-beta Electric Co Ltd to supply and install 2 Nr Stelrad Elite radiators 600 x 900 27 sec K1 complete with thermostatic valves on Fire Exit side wall within refectory for the sum of £429.11

7.02 Please instruct Alpha-beta Electric Co Ltd to supply and install 2 Nr sweep fans mounted in refectory at equal intervals. Switch each unit by a speed controller and protect by lined spur units for the sum of £481.64

For office use: approximate costs
Omit £    Add £

| | | | | Omit £ | Add £ |
|---|---|---|---|---|---|
| ☑ Contractor | ☑ Architect | ☑ M&E Consultant | Amount of contract sum Approximate value of previous instructions | | |
| ☑ Employer | ☑ Clerk of Works | ☑ File | Sub-total Approximate value of this instruction | | |
| ☑ Quantity Surveyor | ☑ Structural Engineer | ☑ Alpha-beta Co Ltd | Approximate adjusted total | | |

*C Miller*

Signature of Contract Administrator
XYZ Group

Figure 1.25(a)   Sample contract instruction form

# Contract instruction

Alterations to
Engineering services

Site address

Description of work

Contractor address

Job reference

Issue date

Instruction no

The contract sum will be adjusted where applicable in accordance with the terms of the relevent contract condition.

Instruction

| | For office use: approximate costs |
| | Omit £ | Add £ |

☐ Contractor  ☐ Architect  ☐ M&E Consultant

☐ Employer  ☐ Clerk of Works  ☐ File

☐ Quantity Surveyor  ☐ Structural Engineer  ☐

| | Omit | Add |
|---|---|---|
| Amount of contract sum Approximate value of previous instructions | | |
| Sub-total Approximate value of this instruction | | |
| Approximate adjusted total | | |

Signature of Contract Administrator

Figure 1.25(b)   Blank contract instruction form

# Selecting and installing wiring systems

## *Objectives*

After working through this chapter you should be able to:

1  *state the meaning of the term 'wiring system';*
2  *describe some important considerations when selecting and installing the following wiring systems:*

- *Non-armoured PVC/LSF insulated cables to BS 6004 and BS 7211*
- *Armoured PVC/LSF/XLPE insulated cables to BS 6346, BS 6724 and BS 5467*
- *MIMS cables to BS 6207 and accessories to BS 6081*
- *Steel conduit to BS 31 and BS 4568 and PVC conduit to BS 4607, BS 4678 and BS 6099, internally wired with single-core, PVC/LSF insulated cables or other specified types of cable*
- *Steel trunking to BS 2989 and PVC trunking to BS 4678, internally wired with single-core, PVC/LSF insulated cables or other specified types of cable*
- *Busbar trunking to BS 5486 Part 2 (IEC 439–2)*
- *Cable tray systems to BS 729 and support systems wired with specified armoured cables and other wiring systems*
- *Optical fibre cables;*

3  *describe the procedure for selecting cables for wiring systems in specified installations.*

# Wiring systems

This chapter reviews and extends some of the common wiring systems discussed in Chapter 3 of the author's *Part 1 Studies: Theory*. It introduces new cable developments and considers in more detail the selection and installation of various wiring systems in compliance with the *16th Edition IEE Wiring Regulations*.

A wiring system is defined in Part 2 of the Wiring Regulations as:

> An assembly made up of cable or busbars and parts which secure and, if necessary, enclose the cable or busbars.

A **cable** is a length of insulated single-core conductor (or multicore conductors) that is either solid or stranded and covered with insulation. The insulation around the conductor not only provides a degree of mechanical protection but also serves to stop current leakage, either to earth or between other conductors in the cable.

With the exception of large power cables which have aluminium conductors, most general purpose cables have conductors made of high-conductivity copper, meeting the requirements of BS 6360 *Conductors in Insulated Cables and Cords*, (see Note below).

The insulation of these cables is mostly PVC (polyvinyl chloride) which has good electrical properties and is resistant to most oil and chemical contaminations. For installations where an outbreak of fire is a serious threat, low smoke and fume (LSF) compound cables to BS 7211 are recommended. You will come across the abbreviation **uPVC** which means unplasticised PVC. It is PVC but the plasticising agents have been removed from the material formulation giving it a high compression strength and high impact resistance as well as the property of self-recovery.

The word busbar is from the Latin word for omnibus meaning 'for all' and describes a conducting material in the form of a bar. This is usually a bare or lightly insulated conductor, made of copper or aluminium or sometimes copper clad aluminium. It is used in switchgear as a live interconnector or as a live conductor in rising mains and overhead supply systems. The conductor is either solid or hollow, and round or rectangular in shape. To be able to withstand very large thermal and electromagnetic stresses without damage, it is essential that the material used in the construction of the busbar provides the best possible mechanical properties. It must have low electrical and thermal resistance, good mechanical strength (tension, compression and shear strength) and also have high resistance to corrosion.

The parts that are be used to secure the wiring system and keep it in position cover a wide range of fixings, terminations and supports. Some examples are shown in Figure 2.1. The parts that enclose the cable/busbar cover a wide range of different types of insulation material, metal screens, armouring and other forms of enclosures.

**Note:** For the same cross-sectional area and temperature, a copper conductor's resistivity is only 66% that of an aluminium conductor and is the more suitable material for general wiring cables. However, copper has a density more than three times that of aluminium, being 8817 kg/m$^3$ compared with 2558 kg/m$^3$ respectively. For this reason, where weight-saving is important in larger size cables, particularly armoured cables, aluminium is the preferred conductor material.

Some common wiring systems in general use today are:

- non-armoured PVC/LSF insulated cables to BS 6004 and BS 7211;
- armoured PVC/LSF/XLPE insulated cables to BS 6346, BS 6724 and BS 5467;
- MIMS cables to BS 6207 and accessories to BS 6081;
- steel conduit to BS 31 and BS 4568 and PVC conduit to BS 4607, BS 4678 and BS 6099, internally wired with single-core, PVC/LSF insulated cables or other specified types of cable;
- steel trunking to BS 2989 and PVC trunking to BS 4678, internal wired with single-core, PVC/LSF insulated cables or other specified types of cable;
- busbar trunking to BS EN 60439–2;
- cable tray systems to BS 729 wired with specified armoured cables and other wiring systems.

## Cable types and accessories

Each wiring system has its own particular advantages and disadvantages. Selection for a specified installation has to consider a number of important factors such as intended purpose, flexibility of use,

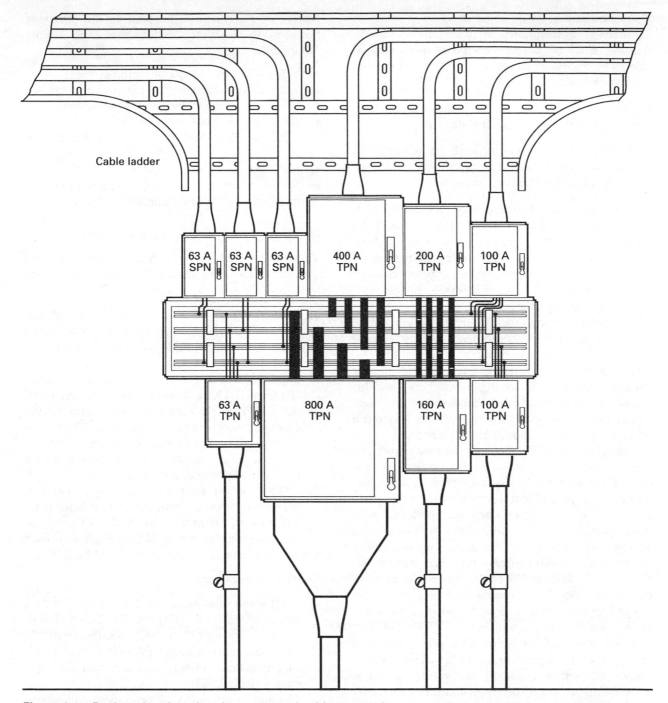

Cable ladder

63 A
SPN

63 A
SPN

63 A
SPN

400 A
TPN

200 A
TPN

100 A
TPN

63 A
TPN

800 A
TPN

160 A
TPN

100 A
TPN

Figure 2.1   Busbar chamber showing armoured cable connections

installation time, durability and appearance. There are other factors and there are many different types of building structure. One must also consider the occupation and use of a building and any environmental condition that might affect the wiring system.

The intended purpose of the wiring system is concerned with information needed to make the correct selection of equipment and wiring materials. The purpose is an ingredient in the assessment of general characteristics which is dealt with in Part 3 of the Wiring Regulations. Any external influences affecting the wiring system will have to be considered and must comply with Regulation 130–08–01. This states that:

All equipment likely to be exposed to weather, corrosive atmospheres or other adverse conditions, shall be so constructed or protected as may be necessary to prevent danger arising from such exposure.

Circuit conductors will need to comply with Regulation 130–02–03 which states that:

All electrical conductors shall be of sufficient size and current-carrying capacity for the purposes for which they are intended.

Premises that have been assessed to cater for high levels of utilisation are often designed around the selection of several wiring systems. Some of these systems, such as trunking, provide sufficient flexibility to allow other wiring systems, such as conduit and MI cable, to be fully integrated. In this way, cost savings in **installation time** allow the wiring to be completed more quickly and more efficiently. For example, a factory premises having a large number of machines in different places of the factory might consider using an overhead busbar trunking system with tapping boxes every 3 m. This allows the machines to be connected to the overhead trunking using other wiring systems which can be easily erected or modified as required.

The term 'durability' is concerned with how long a wiring system will last without significant deterioration through age or exposure from any external influence (e.g. ambient temperature (AA), presence of water (AD) or corrosive, polluting substances (AF)). It is very important that the chosen wiring system is given the necessary **protection** against any harmful affects that may shorten its working life. Section 522 of the Wiring Regulations (see Note below) lists a number of common external influences, others can be found in Appendix 5 of the *IEE Wiring Regulations*.

It is the use of a building that often dictates a wiring system's appearance. In dwellings, shops, offices and other premises the wiring system is selected mainly on aesthetic grounds. It is generally concealed beneath the surface (often plaster) with only the electrical accessories and functional equipment exposed to view. This method is commonly known as 'flush wiring' but where appearance is not important, the wiring system can be fixed directly to the surface, known as 'surface wiring'. The latter is frequently chosen for factories, workplaces and buildings that are not altered for one reason or another.

**Note:** Attention should be paid to the following Sections in Chapter 55 of the *IEE Wiring Regulations*:

- 521 – selection of type of wiring system
- 523 – current-carrying capacity of conductors
- 524 – cross-sectional areas of conductors
- 526 – electrical connections
- 527 – selection and erection to minimise the spread of fire
- 528 – proximity to other services
- 529 – selection and erection in relation to maintainability, including cleaning.

*Non-armoured PVC/LSF insulated cables*

These are a cost-effective type of wiring system. The following cables are available which are shown in Figure 2.2:

1. 450/750 V single-core non-sheathed general purpose cables having cross-sectional areas ranging from 1.5 mm$^2$ to 400 mm$^2$ (Ref. 6491X), used for the internal wiring of luminaires, appliances and control circuits. LSF cables (Ref. 6491B) have limited cross-sectional areas.
2. 300/500 V single core-sheathed cables having cross-sectional areas ranging from 1.0 mm$^2$ to 35 mm$^2$ (Ref. 6181Y) used for internal wiring of equipment and control circuits where added mechanical protection is required.
3. 300/500 V flat twin and three-core sheathed cables having cross-sectional areas ranging from 1.0 mm$^2$ to 16 mm$^2$ (Ref. 6192/3Y) used for general purpose wiring. Also cables with bare circuit protective conductors (Ref.6241/2/3Y).

Other BS cables are:

Single-core heat-resisting PVC non-sheathed cables (Ref.6491X (HR) and Ref.2491X (HR)). Circular twin and 3 or 4 core PVC insulated and sheathed, heat-resisting cables (Ref. 3092/3/4Y). Flat twin and rubber, insulated and sheathed cables (Ref. 6192P). Circular twin and 3, 4 or 5 core vulcanised rubber insulated, tough rubber sheath flexible cords (Ref. 3182/3/4/5.

Some of these cables are available in PCP (polychloroprene) and EPR (ethylene propylene rubber) and may be screened with a copper braid.

The cables used for the wiring of domestic premises, as well as many other types of premises,

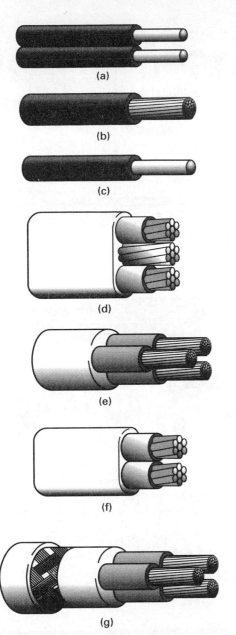

Figure 2.2 Non-armoured insulation cables:
(a) bell wire; (b) single-core non-sheathed flexible
cable; (c) single-core non-sheathed solid cable;
(d) flat twin sheathed cable with bare circuit
protective conductor; (e) circular 3-core PVC
insulated and sheathed, heat-resisting cable;
(f) flat twin rubber insulated and sheathed cable;
(g) circular 3-core PVC insulated sheath screened
cable

are the flat PVC insulated and sheathed cables
incorporating bare circuit protective conductors
(Ref. 6242/3Y). The installation methods for these
cables are illustrated in Table 4A of the Wiring
Regulations (Methods 1 and 3). Where, in Method 1,

they are run beneath or directly on a surface, additional mechanical protection using channelling or
conduit may be required. This is to avoid the risk of
damage through impact (see all requirements under
Regulation 522–06 of the *IEE Wiring Regulations*).

In the process of terminating the cables, it is
important not to remove too much insulation or cut
conductor strands out in an attempt to ease connections into crowded terminals. You should not
overtighten terminal screws as this will not only
ruin the screw but also compress and weaken the
cable conductors. You should always allow sufficient slack cable inside enclosures to make the connections properly and to gain easy access when
carrying out routine inspection and testing. It is
important to identify every conductor by its proper
colour code, as described in Tables 51A and 51B of
the *IEE Wiring Regulations*. You should use heat-
resistant sleeving on cables that are placed in high
ambient temperatures such as found inside luminaires and heating appliances. Table 52B of the
Wiring Regulations shows the maximum conductor
operating temperatures for different material
insulation.

In terms of fixings, you should use the recognised
cable clips and space them as required by Table 4A
of the *IEE On-site Guide* or Table 12 of the *IEE
Guidance Notes, Book No. 1, Selection and
Erection*. In this latter book (Table 11), you can
find the minimum internal radii of cable bends in
Table 11. You should note a factor of 3 for single-
core, circular cables having stranded conductors,
when such cables are installed in conduit, ducting
or trunking.

Clauses in a typical electrical specification covering PVC cables might read as follows.

### PVC cables

All cables shall be installed off drums and not loose
coils.

Single-core cables shall generally be installed in trunking and conduit and care shall be taken to ensure that
strict segregation of circuits for various services is
maintained as required by the *IEE Wiring Regulations*.

Where conductors are terminated into screw or clamp-
type terminals, solid conductors shall be bent back to
form a double thickness and stranded conductors shall
be twisted. Where more than one conductor terminates
in a single terminal, the cores of different conductors
shall not be twisted together.

Where conductors are terminated by use of crimped-
type cable terminations, the type and size of the crimp
terminal shall be in accordance with the manufacturers'

recommendations and crimps shall be applied using the recommended crimping tool.

Cables passing through luminaires shall be of the heat-resisting type.

All cable cores shall be of the correct colour coding and where twin cables with bare CPCs are used for switch connections, both live conductors (switch feed and switch wire) are to be insulated with red PVC. The bare CPC shall be insulated with green/yellow PVC sleeving. All conductors shall be correctly terminated.

A separate CPC shall be connected between the earth terminal of accessories and the earth terminal of metal boxes.

Jointing of cables shall not be permitted unless specifically requested. Where allowed, the joints shall be installed in accessible locations in joint boxes to BS 6220 or in purpose-made fully-enclosed terminal boxes with fixed terminals. In all cases, terminal boxes shall be securely fixed.

Runs of flat PVC/PVC cables embedded in walls should meet the requirements of Regulation 522–06–06 of the *IEE Wiring Regulations*.

### Flexible cables and cords

These shall be 450/750 V grade with stranded copper conductors, complying with BS 6500.

Flexible cables where used in ambient temperatures not exceeding 60°C shall be multicore PVC insulated and PVC sheathed.

Final connections to luminaires shall be multicore EPR insulated and CSP sheathed cable complying with BS 6007.

All flexible cords shall not be less than 0.75 mm² and where the temperature exceeds 60°C the following types of cable should be used:

not exceeding 85°C heat resistant PVC sheath or or rubber HOFR sheath;
not exceeding 150°C silicon-rubber insulated and braided;
not exceeding 185°C glass fibre insulated braided circular.

*Armoured PVC/LSF insulated cables*

These cables (Figure 2.3) have considerable use in industry for all kinds of application. They are also extensively used to provide electricity supplies to consumers' premises as well as temporary supplies on construction sites. Some types of cable in the low voltage range are as follows:

- 600/1000 V single core cables with circular, stranded copper conductors of sizes ranging from 50 mm² to 1000 mm² (see Note below).
- 600/1000 V multicore cables with shaped,

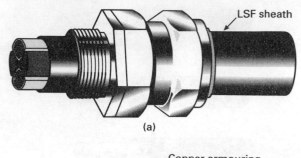

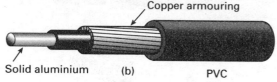

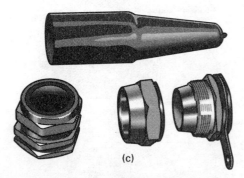

Figure 2.3   Armoured cable and accessories: (a) 4-core XLPE insulated SWA, LSF-sheathed cable; (b) single-core PVC concentric service cable; (c) armoured cable glands and accessories

stranded copper conductors of sizes ranging from 1.5 mm² to 400 mm² (see Note below).
- 600/1000 V PVC concentric service cables with circular, solid/stranded, copper/aluminium conductors from 4 mm² to 50 mm²

**Note:** Single core and multicore cables of similar ratings as above are available with aluminium conductors.

The armouring and outer sheath of these cables provide good quality protection against a number of external influences. Some types of cable (not necessarily armoured) are designed with thermosetting insulation. These are called **crosslinked polyethylene (XLPE)** insulated cables designed to BS 6899. They have a maximum continuous operating temperature of 90°C compared with 70°C with ordinary PVC insulated cables. This advantage allows XLPE cables to operate with higher current carrying capacities and improved short-circuit performance. Where voltage drop is not a problem, smaller size cables by comparison to PVC may be used.

Table 4A of the Wiring Regulations shows several installation methods for armoured cables. It is important that proper lifting gear is used when unwinding heavy cable drums and gloves and protective footwear should be worn by operatives. Cable manufacturers provide information to users about removing steel binding straps and wooden battens from the drums. The battens should be stored in a safe place and any protruding nails removed or bent back. *HSE Guidance Notes, PM 15 Safety in the use of timber pallets, 1993* is worth reading.

With the exception of LSF cables, it is recommended that cables with PVC insulation are handled only when their temperature and the ambient temperature are above 0°C for at least 24 hours. The minimum bending radii and maximum spacing of clips for armoured cables is given in Tables 11 and 12 respectively of the *IEE Guidance Notes, Book No. 1 Selection and Erection.*

Clauses in a typical electrical specification covering PVC armoured cables might read as follows.

### Armoured cables

All cables shall be installed in complete lengths, and, unless specified, no 'through' joints are allowed.

Cables shall be fixed by means of proprietary cable clips, cleats or clamps mounted on suitable channel support systems.

The fixing centres for all cable supports shall be as indicated in the current edition of the *IEE Wiring Regulations.*

All cables shall be terminated by means of purpose-made brass glands and be of the type that clamps both the wire armouring and the PVC outer covering. The glands shall incorporate earthing tags and be fitted with PVC shrouds.

'Tee' joints, and where specified, through joints shall be made within a cast iron mould incorporating internal armour clamps filled with a hot pouring compound or a plastic mould filled with a cold pouring compound.

Joints shall be made using hot tinned copper jointing ferrules or crimp-type ferrules.

### MIMS cables

Mineral insulated metal sheath cables, often referred to as MICC cables since the outer sheath is made of copper, although costly, have a number of advantages over other types of cable. They are fireproof, resistant to corrosion and have good mechanical strength (Figure 2.4). The magnesium oxide insulant and the copper conductors ensure that the cables have a long life expectancy. The following cables are available:

- 500 V light-duty, 1.0 mm² to 4.0 mm² single or multicore cables
- 750 V heavy-duty, 1.5 mm² to 240 mm² single or multicore cables.

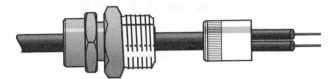

Figure 2.4   Mineral insulated cable termination

**Note:** Multicore cables are restricted by size.

The cables are relatively easy to install and find considerable use in Category 3 circuits for fire alarms and emergency lighting. They also find use in food processing plants, boiler rooms, garage forecourts, hostile environments and safety control circuits. In this latter application, where electromagnetic system interference is a problem, a special twisted conductor cable is available.

Although the *IEE Wiring Regulations* cable tables show a permitted sheath operating temperature of 70°C or 105°C, MI cables are capable of operating indefinately at 250°C. They are not recommended to be used in circuits supplying inductive loads that create transient voltages. Where this is impossible to overcome, surge diverters need to be installed. For added protection when installed underground or exposed to the weather or risk of corrosion, the copper sheath is provided with an overall PVC covering (LSF for fire protection).

Table 4A of the *IEE Wiring Regulations* shows several installation methods for these cables. It is important that the minimum radii of any cable bend and the maximum spacing of clips complies with the values given in Tables 11 and 12 respectively of the *IEE Guidance Notes, Book No 1, Selection and Erection.*

It is recommended that the termination procedures for these cables is carried out using the proper termination tools and materials. Cable end seals and glands are manufactured in accordance with the requirements of BS 6081.

Clauses in a typical electrical specification covering MICC cables might read as follows.

## MICC cables

All cables shall be heavy-duty fixed by means of single or multi-way copper saddles, purpose-made by the manufacturer of the cable. Where used for the fixing of PVC sheathed cables the saddles shall also be PVC sheathed to match.

The fixing centre of saddles for bare cables shall not exceed the spacings indicated below:

| | | |
|---|---|---|
| 1.5 mm² to 2.5 mm² | 600 mm (Hor) and 800 (Ver) |
| 4 mm² | 900 mm (Hor) and 1200 (Ver) |

At bends, sets, boxes, fittings etc. cables shall be additionally supported by saddles at each side where necessary.

The minimum bending radius shall be not less than six times the overall diameter of the particular cable.

### Steel conduit and single core insulated cables

This wiring system can be used as a 'flush' or 'surface' form of wiring with the conduit providing a high degree of mechanical protection and circuit wiring adaptability.

The conduit can be welded or solid drawn and made in metric sizes of 16 mm, 20 mm, 25 mm and 32 mm. Each length is 3.75 m and in the latter three sizes there are normally eight lengths per bundle.

Conduit is classified according to its method of assembly and its protection against corrosion. The two most common classifications of finish are Class 2 and Class 4. The former is a stove enamel black finish which offers medium protection for inside and outside use and the latter is a hot dipped galvanised finish which offers high protection for inside and outside use. A range of different conduit fittings and tools are available (see *Part 1 Studies: Practical* page 37).

It is important for conduit systems to be erected and fixed in position before any wiring is installed (see Regulation 522–08–02 of the *IEE Wiring Regulations*). You should pay particular attention to any environmental condition which might cause corrosion (see the requirements under group Regulation 522–05 of the Wiring Regulations and also page 71 of the *IEE On-site Guide: Protection against corrosion of exposed metalwork or wiring systems*).

When wiring between buildings, you should know the maximum length of span for conduit is 3 m and the minimum height above ground is 5.8 m at road crossings. This can be reduced to 3 m in positions inaccessible to vehicular traffic but not in agricultural premises. The spacing of supports for conduits can be found in Table 4A of the Site Guide or in Table 14 of the *IEE Guidance Notes, Book No.1, Selection and Erection*. No supports are normally needed for non-sheathed cables inside conduit but in vertical runs exceeding 5 m there is a possibility that some mechanical stress, mainly compression, may occur.

### PVC conduit and single-core PVC insulated cables

Much of what has already been said about metal conduit applies to plastic conduit. However, the system does not provide the same degree of mechanical protection and may be subject to distortion if fixings and supports are incorrect. Plastic conduits are considerably robust and corrosion resistant and they can be installed relatively quickly, having the same degree of flexibility as metal conduit. Manufacturers make round unscrewed conduit in common outside diameter sizes of 16, 20, 25, 32, 38 and 50 mm. The conduits, black or white in colour are in lengths of 4 m and can be light guage (LG) or heavy guage (HG) having plain ends. Light guage conduit is normally used for surface work and heavy guage suitable for concealed work and in floor screeds. Joints are formed using a sealing cement.

Like metal conduit a wide range of accessories are available including heat resistant accessories. Circular boxes are designed to suspend loads up to 3 kg (or 10 kg with external lugs) at 60°C. Manufacturers make oval conduits and channelling for use in plastered walls and they also make flexible conduits for final connection to movable or vibrating equipment.

In terms of installation and erection, manufacturers recommend that sizes up to 25 mm diameter can be bent 'cold' using the correct size bending spring. These are often colour coded (green for HG conduit and red/white for LG conduit) and the springs should be in good condition. In very cold weather, it is recommended that the conduits are warmed slightly otherwise there is a tendency for them to kink or fracture during the bending process. Heating ('hot' bending) is the normal practice for conduit in sizes above 25 mm in diameter. Where there is a possibility of expansion, expansion couplings are to be used at intervals of 4 m (or less). For spacing of clips see Table 4C of the IEE on-site guide. It is important to install a separate protective conductor in these conduit systems.

Clauses in a typical electrical specification covering metal conduit might read as follows.

## Conduits and accessories

*Steel Conduit.* All steel conduit shall be heavy gauge 'screwed' welded and not less than 20 mm diameter, unless otherwise specified.

All conduits and accessories shall fully comply with BS 4568, Parts I and II and BS 31 and shall be provided by a manufacturer having a licence to use the British Standards Institution mark.

Conduits and accessories shall be Class 2 finished black enamel, or Class 4, hot dipped galvanised and conduit boxes shall be the standard small circular type and be fitted with malleable iron or pressed steel covers secured by means of brass roundhead screws. All conduit boxes shall be provided with tapped spout entries.

Where adaptable boxes are used, they shall be the pressed steel or malleable iron type and shall be of minimum size 100 × 100 × 50 mm.

The routes of all conduits and the location of all points and draw-in positions shall be to the approval of the Architect or his appointed representative.

Under no circumstances shall cables be drawn into the conduit until all such conduit, bends, boxes or other fittings have been permanently fixed in position and approved by the Architect or his appointed representative, and all associated concreting and plastering work completed.

The separate conductors of the same circuit shall in all cases be drawn into one conduit only. All conductors within a conduit shall be drawn in at the same time.

Cables forming final circuits connected to different distribution boards shall not be drawn into the same conduit.

If for any reason it is necessary to fit draw-in boxes in conduit runs, such boxes shall be made permanently accessible and so arranged to be neatly finished and flush with the finished surface of walls, floors and ceilings. Draw-in boxes shall be installed on all conduit runs every 10 m of a straight run and with not more than two right angle bends, except where specified.

The conduit system must be mechanically and electrically continuous throughout, so that the cables are fully protected.

Adequate precautions shall be taken by the Contractor to prevent the ingress of silt or moisture into the conduit system, but should this occur the conduit shall be swabbed clean before the wiring commences.

Where conduits are run in cavity walls and in all situations exposed to dampness, the conduit and conduit accessories shall be of Class 4 finish.

Surface-mounted metal conduit shall be secured by spacer-bar saddles with fixing centres not exceeding the following sizes:

| | |
|---|---|
| 25 mm dia | 1750 mm (Hor) and 2000 mm (Ver) |
| 25 mm–35 mm dia | 2000 mm (Hor) and 2250 mm (Ver) |

All saddles and boxes shall be securely fixed to the surface by screws of adequate size and length.

Flush-mounted conduit shall be installed in suitable chases cut into the building fabric, to a depth such that a minimum of 12 mm cover to the finished building surface is attained. Flush conduits shall be fixed using pipe hooks or crampets driven into the building structure.

Where conduit paint finish is damaged in erection, it shall be made good, together with exposed threads, in black enamel or zinc rich paint as appropriate.

Where conduit is run over a structural expansion joint, an approved conduit expansion coupler shall be installed, complete with external protective conductor connected across the conduit expansion joint.

*Flexible conduit.* Flexible steel conduit to BS 31 shall be manufactured of helically wound zinc plated steel, with string packing, and sheathed with black PVC. (See Figure 2.5)

The interior surface of the flexible conduit shall be free from burrs and sharp edges, and each length shall be provided with a proprietary type of screwed brass male adaptor at each end. A protective conductor shall be provided for all conduit lengths.

## Trunking and specified types of cable

This wiring system, like conduit, is a protective enclosure system for cables and may be of metal or plastic construction in heavy duty or light duty material. Trunking systems are often designed for a special purpose, such as lighting trunking, dado/skirting trunking, bench trunking, floor/underfloor trunking. It can be designed to suit environmental conditions (weatherproof) and for the wiring of category circuits, as compartmental trunking to provide facilities for low current, telephone, power and data cables.

Standard metal trunking for surface wiring can

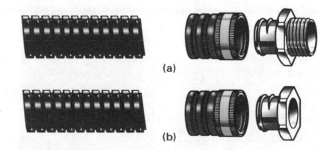

(a)

(b)

Figure 2.5   Types of flexible conduit connection: (a) for attachment into a threaded entry or insertion into plain holes using a locknut; (b) for attachment to enclosures with plain holes where the insert not only locks the conduit but also acts as a smooth bush

be manufactured with different finishes, such as electroplated zinc sheet steel complying with BS 1706 and finished in a stove enamel dark, admiralty grey or a hot dipped galvanised sheet steel complying with BS 2989, or a prime bright spangled galvanised steel finish complying with BS 4678. A full range of sizes is given in Table A7 of the *IEE Guidance Notes No. 1 Selection and Erection.*

Lengths for special systems vary but standard lengths are normally 2 m or 3 m, up to depths of 200 mm. Lid fixings vary according to manufacturers' preferences and a varied range of accessories are available. Reference should be made to the *IEE Wiring Regulations, Table 4A Schedule of methods of installation of cables,* and also the relevant regulations contained in Sections 521 to 529.

Cable capacities for trunking systems are found in Appendix 5 of the *IEE On-site Guide* and Appendix A of the *IEE Guidance Notes No. 1 Selection and Erection.*

Spacing and supports for trunking are given in Table 15 of the guidance notes.

Much of the above refers to plastic trunking but like plastic conduit, separate protective conductors are required. The small section trunkings like slimline trunking, miniature trunking and skirting trunking, find considerable use in offices and commercial premises to provide computer and data supplies as well as small power and security requirements. A typical trunking layout is shown in Figure 2.6.

Clauses in a typical electrical specification covering metal trunking might read as follows.

### Metal trunking

Trunking shall be manufactured to BS 4678, Part I 1971 from mild sheet steel complete with returned edge for secure lid fixing by means of screws at no more than 900 mm centres. The trunking shall be of the following guages:

| | |
|---|---|
| 50 mm × 50 mm | 20 SWG (1.0 mm) |
| 100 mm × 150 mm | 18 SWG (1.2 mm) |
| 150 mm × 150 mm | 16 SWG (1.6 mm) |

The total sectional area of cables installed shall be in accordance with the current IEE Wiring Regulations and shall not exceed 45% of the internal c.s.a. of the trunking.

Where cutting or slotting of the trunking is carried out, all sharp edges and burrs shall be removed. Where cables pass through slots in the trunking the perimeter of the slots shall be shrouded with continuous PVC sleeving or similar insulated sleeving.

Cables installed in the trunking shall, wherever practicable, be laid with the larger size of cables at the bottom.

Where trunking is installed with the open side down, suitable cable retaining straps shall be installed in each compartment of the trunking at a maximum spacing of 900 mm.

To ensure effective earth continuity, each joint in the trunking (between lengths, tee and angle units, etc) requires the connection of a copper bonding link of approximate size 38 × 13 × 1.6 mm). The contact surfaces shall be thoroughly cleaned.

Where conduit is terminated in trunking, the trunking shall be scraped clean of any paint and primer.

### *Busbar trunking*

These systems are designed for a.c. loads ranging between 40 A and 4500 A. For light industrial applications, ratings up to 80 A are typical. The trunking is usually in 3 m lengths with facilities for

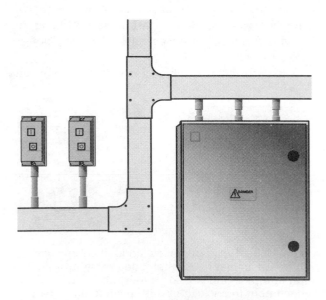

Figure 2.6   Typical layout of PVC heavy-duty trunking system

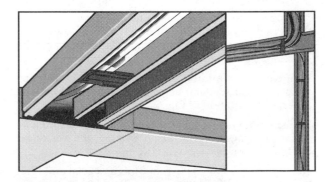

Figure 2.7   Cable retainers in trunking

plug-in tapping boxes at 0.3 m intervals. The tapping boxes of 32 A rating are MCBs, fuses, push button motor protection or socket outlets and provide the final circuit protection/isolation. They are clipped to the busbar system which has sliding shutters to close the outlets when not in use.

For factory power distribution and rising mains, busbar ratings up to 450 A are satisfactory and the tapping boxes contain fuses, MCBs or fuse switches of ratings 32 A, 63 A, 100 A and 200 A. In higher rated systems the tapping boxes will contain HBC fuses and MCCBs.

From the point of view of expansion under varying ambient temperature conditions, the joint links between the standard lengths of less than 250 A rating can satisfactorily absorb movement but in higher ratings where the joints are rigid and runs exceed 30 m, expansion fittings, in the form of copper braided flexible connections, are required.

Section 527 of the *IEE Wiring Regulations* is concerned with the selection and erection to minimise the spread of fire and group regulations 527–02 is concerned with sealing the wiring systems for this purpose. The arrangements for sealing are explained in Regulation 527–02–03. Fire barriers are required where the trunking passes through a fire rated floor or wall and has to be made from an approved fire-resisting material such as rock-wool or expanding compound which seals the internal space of the enclosure. For further information see *IEE Guidance Notes No.4, Section 3, item 3–3–2, Spread of fire*.

Clauses in a typical electrical specification covering busbar trunking such as in Figure 2.8 might read as follows.

### Busbar trunking

The busbar trunking shall be manufactured to comply with BS 5486 Part 2 (IEC 439–2) and as detailed in drawings/schedules complying to IP42 or IP20 as specified in BS 5420 (IEC 144).

The trunking enclosure shall be made from shaped extruded aluminium side members with a total c.s.a. of 718 mm². The back panels shall be of 18 SWG galvanised sheet steel and the front modular covers also 18 SWG, but made from zinc coated sheet steel. The paint finish being to BS 591 orange unless otherwise specified.

The trunking conductors shall be rectangular section copper, aluminium or cuponal as specified in the drawings/schedules. The conductor bars shall be sleeved throughout their length in phase coloured PVC and supported at 300–600 mm intervals. The support insulators are to be manufactured from extruded aluminium cross-members with moulded insulator inserts.

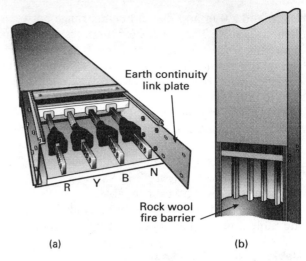

Figure 2.8   Busbar trunking: (a) view showing side member to provide earth continuity and fully prepared conductor bars; (b) section removed to show fire barrier

Fire-resisting barriers shall be provided with the trunking where the enclosure passes through fire division walls, floors, ceilings or partitions.

Where a change of direction, termination or rating occurs, the specified manufacturers' purpose-made fittings shall be used.

Tapping boxes shall be of the clamp-on type and fitted with HBC fuses, MCCBs or switch fuses as detailed in the drawings/schedules. The boxes shall be provided complete with integral flexible cables terminating in purpose-made shrouded brass connection clamps for the attachment to the busbars without the need for drilling.

Where there is a risk of expansion, due to high ambient temperatures, copper braid conductors shall be fitted in the trunking in accordance with the manufacturers' recommendations.

### Metal cable trays and cable ladders

Metallic cable trays are used mainly as a means of supporting armoured cables. Standard lengths are normally 3 m but variations occur with some manufacturers. Trays are made from mild steel or stainless steel and comply with BS 1449 having standard hot-dipped galvanised steel finish, and BS 729 with an admiralty-type perforation pattern. They are manufactured to BS 2989 (pre-galvanised steel) or BS 2989 (post-galvanised steel).

Light duty trays have a gauge (thickness) between 1.0 mm and 2.0 mm. The width measurement is between 50 mm and 900 mm and the height of the L-shaped flange varies between 12 mm and 20 mm. Heavy-duty trays have gauges between

1.5 m and 2.0 m and flange heights range between 38 m and 51 m. The flanges for these trays are bent over. They are called return flanges and are more robust, giving extra rigidity. They are available in width sizes up to 610 m. Figure 2.9 shows a selection of perforated cable trays and some accessories.

Cable ladder systems are generally designed for heavy cable distribution and pipes in factories, power stations, steel works, etc. The ladders are normally manufactured in standard lengths of 3 m and vary in width. Figure 2.10 shows a number of different shapes that are available.

Clauses in a typical electrical specification covering cable trays such as in Figure 2.10 read as follows.

**Cable trays and ladders**

These shall be installed as required and shall be of the admiralty pattern, medium- or heavy-duty return flange type or as otherwise specified.

The trays shall be coupled by means of rigid edge bars and fish plates, and care shall be taken to eliminate dangerous bolt ends projecting through the flanges.

Effective earth continuity between sections of the tray shall be ensured by separate copper bonding strip bolted across each coupling point. The contact surfaces shall be thoroughly cleaned prior to bonding.

Full use shall be made of standard auxiliary components or bends, tees, reducers, intersections and risers to suit the installation.

Cable trays for horizontal runs, suspended from the ceiling shall be installed with the flanges facing upwards, and the trays shall be supported with

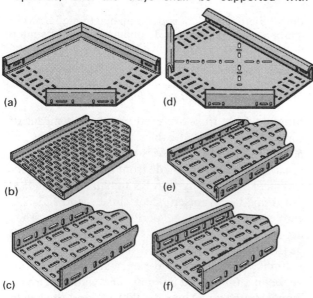

Figure 2.9   Types of cable tray: (a) heavy-duty 90° flat bend; (b) light-duty; (c) heavy-duty; (d) medium return flat tee; (e) medium return; (f) heavy return

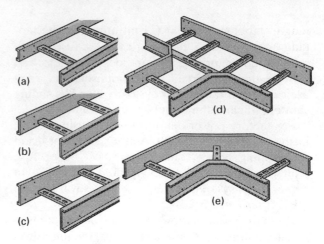

Figure 2.10   Types of cable ladder: (a) light-duty; (b) medium-duty; (c) heavy-duty; (d) flat tee; (e) flat 90° bend

purpose-made mild steel trapeze straps or brackets at sufficient centres to prevent the tray from sagging.

The complete installation shall be free from burrs and sharp edges.

Cables associated with telephones, microphones, radio, and television systems shall not be run on a common tray with any other services.

*Optical fibre cables*

These are communication cables for use mainly in computing systems and data transmission systems. They find considerable application in structured cabling systems in modern office buildings. One electrical installation application is in emergency lighting systems where light from a halogen lamp projector box is made to pass down a series of fibre optic cables terminated in the ceiling. The light output must comply with BS 5266. Some of the features of fibre optic cables are:

- they have immunity from electromagnetic interference
- they do not radiate electromagnetic energy
- signals can be transmitted at high speeds over long distances
- they are intrinsically safe and if cut, there is no risk of a spark to cause explosion
- the cables can be provided with an LSF sheath for additional fire safety
- they are easy to install and can replace whole looms of copper cable

Briefly, an optical fibre consists of a silica core and a silica cladding. Light moves down the core by means of total internal reflection at the core/

cladding interface. It is the ratio of the core/cladding that is used to describe the various sizes of fibre that are used (e.g. 50/125, 62.5/125, 85/125, 100/140).

The optical characteristics of the fibres are concerned with **attenuation** (loss of power), **bandwidth** (information carrying capacity) and **numerical aperture** (measurement of angular light acceptance). These will not be discussed, but the design of the cables fall into two basic categories, namely:

(i) light jacket cables;
(ii) loose tube cables.

Jacket cables are normally designed for indoor use and consists of 0.9 mm tight buffered fibres in a special reinforced polyamide fibre annulus.

The cables are enclosed in an outer polymer jacket or sheath and are often called simplex cables (single fibre), duplex cables (two simplex fibres together) or multicore cables (4–24 fibres). Both simplex and duplex cables are used for point-to-point data link connections or pre-terminated patch/jumper cord connections. The multicore cables are generally used as backbone cables in local area networks (LAN), direct on-site connections within protective enclosures or ruggedised pigtails.

Loose tube cables while similar in design to tight jacket cables, come in a wide variety of configurations, both in number of fibres and construction. A typical LSF, waterproof cable consists of individually coloured primary coated fibres contained within a gel-filled polyester tube. Up to 12 such tubes are stranded around a non-metallic strength member. The internal spaces are filled with an LSF compound and a black polyethylene inner sheath forms a moisture barrier. The cable is finally covered with an LSF black outer sheath. Loose tube cables require some form of splicing tool to make the necessary termination. This is often a cleaving tool or fusion splicer.

Figures 2.11 and 2.12 shows the construction of fibre optic cables. Manufacturers produce a wide range of termination kits and accessories as well as test equipment.

Clauses in a typical electrical specification covering the cables might read as follows.

### Fibre optic cables

These cables shall be installed where specified in the drawings/schedules.

The system designer shall calculate a typical loss bud-

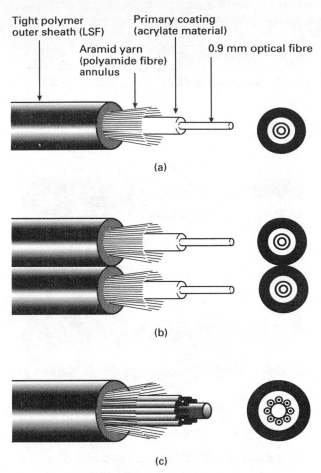

Figure 2.11 Tight jacket optic fibre cables: (a) ruggedised simplex cable; (b) ruggedised duplex cable; (c) internal distribution cable

get for each system installed in order to provide a specification against which the completed installation should be compared.

The cable shall be installed taking the necessary precaution, using proper termination techniques and recognised connectors.

It is essential that exposed fibres are protected in suitable enclosures, such as joint boxes and breakout boxes.

On completion of every circuit the cables shall be tested for their attenuation (in dB) and any possible unsatisfactory spliced joints. For this purpose a proprietary power meter and light source, and OTDR (optical time domain reflectometer) shall be used.

## EXERCISE 2.1

1. Describe the proper termination procedures for the following types of cable/wiring system, stating the materials and tools used.
   a) A 3 m length of flat twin 2.5 mm² rubber

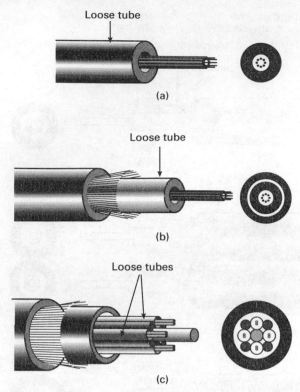

**Figure 2.12** Loose tube fibre optic cables: (a) internal dry cable; (b) external/filled duct cable; (c) heavy-duty duct cable

insulated PCP sheathed cable complying with BS 6007 (Ref. 6192P), terminated inside a BS 490 black PVC adaptable box via a 20 mm plastic cable gland ('Walsall' CP1273). The cable and box are fixed to a wooden surface. The cable runs horizontally for 2 m and then is bent at right angles to run vertically for 1 m and then terminated.

b) A 4 m length of 25 mm² armoured 3-core XLPE insulated cable having shaped solid aluminium conductors, complying with BS 5467, run on an existing length of heavy-duty steel cable tray. The tray is bent at right angles using a proprietary 90° bend and the cable is clipped around the bend and terminated in the top of a 100 A TPN metalclad switchfuse.

c) A short length of Class 4 galvanised steel conduit of size 25 mm and a short length of 25 mm flexible conduit ('Kopex' LT–S, FLH05), terminated in a weatherproof through box ('Walsall F522').

## Cable selection

This topic has already been dealt with in *Electrical Installation Part 1: Theory*. Having outlined several common wiring systems, the intention here is to show how cables are selected for circuits in specified installation conditions. The procedure is as follows:

1.  Determine the circuit **design current** $(I_b)$. This may be stated on load equipment or it may need calculating. In the latter case, information is required of declared voltage $(U_0)$ and type of supply and phases (e.g. a.c./d.c., 1–ph/3–ph). It might also consider power factor and any diversity allowance (see Appendix 1 of the *IEE On-site Guide*).

2.  Determine the current rating/setting of each circuit **overcurrent protective device** $(I_n)$. These devices are commonly known as BS 88 fuses, BS 1361 fuses, BS 1362 fuses, BS 3036 fuses, BS 3871 m.c.b's (Type 1, 2, 3, B, C and D) and also BS 4752 m.c.c.b's (normally used in circuits operating in excess of 100 A). Tables and operating characteristics of some of these devices can be found in Appendix 3 of the *IEE Wiring Regulations*.

3.  Determine the **tabulated current carrying capacity** $(I_t)$. You should refer to Item 6 of Appendix 4 in the Wiring Regulations which deals with determination of size of cable to use and lists appropriate correction factors that may need to be applied to single circuits and groups. It is important to note that the Tables of current carrying capacities (Tables 4D1 to 4LA) are based on an ambient temperature of 30°C. Any variation to this temperature requires a correction factor to be used, see Regs tables 4C1 and 4C2. Grouping correction factors are found in Regs tables 4B1 and 4B2. Others correction factors apply to the use of a BS 3036 fuse (0.725) and bare MI cables exposed to touch (0.9).

4.  Determine the **voltage drop** *(V)* in the circuit . This has to be done in accordance with the requirements of group Regulation 521–01 of the Wiring Regulations. Once the voltage drop is found, it must be checked against the stated allowance.

5.  Determine the actual value of **earth fault loop impedance** $(Z_s)$ in the system. This is done to see if the circuit protective device operates within the required times as indicated in Regulation

413–02-08 and Table 41A or where otherwise specified in the *IEE Wiring Regulations*. The maximum values are given in tables 41B1, 41B2, 41C and 41D and where specified in the Regulations.

6. Check the size of the **circuit protective conductor (S)**. This is to verify that if it complies with group Regulation 543–01, particularly the formula in Regulation 543–01–03 or the Table in Regulation 543–01–04.

## Summary of formulae to use

*Circuit design current ($I_b$)*

For single-phase circuits:

$$I_b = P/U_0$$
$$\text{or } I_b = P/(U_0 \times \text{p.f.}) \text{ with power factor}$$
$$\text{or } I_b = (P \times 1.8)/U_0 \text{ with discharge lamps}$$
$$\text{or } I_b = (P \times \text{DF})/U_0 \text{ with diversity factor}$$

For three-phase circuits:

$$I_b = P/(\sqrt{3} \times U \times \text{p.f.})$$

where $U_0$ is the phase voltage to earth (230 V)
$U$ is the voltage between lines (400 V).

*Tabulated current carrying capacity ($I_t$)*

For single circuits

$$I_t \geq I_n/(C_a \times C_i \times C_r)$$

Where $C_a$ is the ambient temperature
$C_i$ is the thermal insulation
$C_r$ is 0.725 for a BS 3036 fuse.

For group circuits

$$I_t \geq I_n/(C_a \times C_i \times C_r \times C_g)$$

where $C_g$ is grouping factor.

**Note:** Item 6.4 in Appendix 4 of the Wiring Regulations allows only one correction factor to be applied to a cable route if it represents the most onerous conditions encountered along the route.

*Voltage drop (V)*

$$V = L \times I_b \times \text{mV/A/m} \quad \text{(actual)}$$
$$V = 2.5\% \times U_0 \quad \text{(maximum)}$$

where $L$ is the length of run
mV/A/m is the millivolt drop per metre.

**Note:** mV/A/m = mΩ/m (i.e. the resistance/metre).

*Earth fault loop impedance ($Z_s$)*

$$Z_s = Z_e + R_1 + R_2 \quad \text{(actual)}$$
$$\text{since} \quad Z_s = U_0/I_f$$
$$\text{and} \quad I_f = U_0/Z_s$$

where $Z_e$ is the external earth fault loop impedance
$R_1$ is the phase conductor resistance
$R_2$ is the protective conductor resistance
$I_f$ is the earth fault current.

*Protective conductor (S)*

$$S = \frac{\sqrt{I^2 t}}{k}$$

where $t$ is the fuse operating time
$k$ is the conductor/insulating material factor.

The following examples serve to show the procedures for cable selection. It is assumed that all overcurrent protective devices provide both overload and short circuit protection.

 *Example 1*

In a domestic premises, a PVC insulated and sheathed cable (copper conductors) complying with BS 6004 (Ref 6242Y) is chosen for a 7.2 kW/240 V bathroom shower. The circuit requires 8 m of cable and for most of the run it is clipped and covered on one side by thermal insulation. The route is shown in Figure 2.13. The following information is made available:

- earthing system is TN–C–S
- external earth fault loop impedance $Z_e = 0.35\ \Omega$
- prospective short circuit current $I_{pscc} = 16$ kA
- BS 1361 fuse protection
- ambient temperature is 30°C
- no diversity allowance applies
- voltage drop must not exceed 4% of the declared voltage (assume 240 V).

Determine a suitable size cable for the circuit that fulfils the requirements of the *IEE Wiring Regulations*.

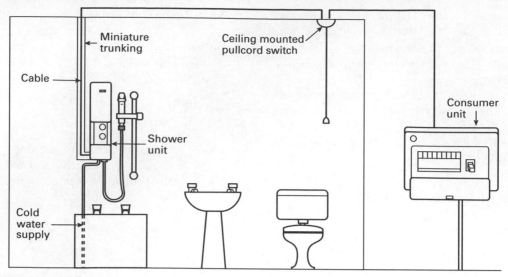

Figure 2.13    Shower installation

*Solution*

The design current of the circuit is found from the formula:

$$I_b = P/U_0 = 7200/230 = 31.3 \text{ A}$$

From Table 41B1 a 30 A BS 1361 fuse is selected having a maximum earth fault loop impedance ($Z_s$) of 1.2 Ω.

With the information provided, the only correction factor that needs applying is thermal insulation. This is considered in Regulation 523–04. You need to consult Tables 4A and 4D2A. In the latter table, under Reference Method 4, you will find in columns 1 and 2 that a 6 mm² cable is rated to carry 32 A This is higher than the design current and the rating of the protective device, satisfying Regulation 433–02–01 ($I_b \leq I_n \leq I_z$).

Now you need to check the voltage drop to see that it does not exceed 9.6 V (i.e. 4% × $U_0$ = 04.40 × 240 = 9.6 V). In Table 4D2B you will find that the voltage drop of a 6 mm² cable is 7.3 mV/A/m. For the length of 8 m and carrying a design current of 30 A, the cable's voltage drop is found by the formula:

$$V = L \times I_b \times \text{mV/A/m}$$
$$= 8 \times 30 \times 7.3$$
$$= 1752 \text{ mV or } 1.752 \text{ V}$$

In order to satisfy the disconnection time of the fuse (0.4 s ), since the circuit is feeding a bathroom shower, the actual circuit impedance has to be found. This can only be done if you know the size of the bare c.p.c. In Table 9A of the *IEE On-site Guide*, three sizes of protective conductor are given. By choosing a c.p.c. of 2.5 mm² the resistances of the phase conductor ($R_1$) and c.p.c. ($R_2$) are: $R_1 + R_2$ = 10.49 mΩ/m = 0.01049 Ω/m. For the stated conditions, length ($L$) being 8 m, external impedance ($Z_e$) being 0.35 Ω and using the 1.38 multipler ($M$) (see Table 9B of the on-site guide), actual value of $Z_s$ can be found. Hence,

$$Z_s = Z_e + (R_1 + R_2) L \times M$$
$$= 0.35 + (0.01049 \times 8) \times 1.38$$
$$= 0.46 \text{ Ω}$$

This value is less than the $Z_s$ maximum value of 1.2 Ω to operate the 30 A fuse in less than 0.4 s and therefore satisfies shock protection.

By choosing a 6 mm² cable with a 2.5 mm² c.p.c., a check must be made on the c.p.c. in case a fault current causes harm to the cable (remember the c.p.c. is a bare conductor and will generate a lot of heat under short circuit conditions). The calculation is made with the adiabatic equation given in Reg 543–01–03 of the *IEE Wiring Regulations*, i.e. $S = \sqrt{I^2 t}/k$. The fault current ($I_F$) is found from the formula $I_F = U_0/Z_s$

(remembering that $Z_s$ is the actual circuit value). Hence:

$I_F = 240/0.46 = 522$ A

If you now consult Figure 1, Appendix 3 of the *IEE Wiring Regulations* (time/current characteristics) you will see that the 30 A BS 1361 fuse will disconnect the supply in a very quick time (approx 0.015 s). Since the $k$ factor for a copper conductor material is 115 (see Table 54C of the Regulations), the size of the c.p.c. can be checked. Thus:

$$S = \sqrt{I^2\ t/k}$$
$$= (\sqrt{522 \times 522 \times 0.015})\ /115$$
$$= 055\ \text{mm}^2$$

This size is less than the chosen c.p.c. size of 2.5 mm² and means that the c.p.c. will safely dissipate the heat from the fault current without damage to the cable. Now try this question with a supply voltage of 230 V, using a 32 A Type 2 mcb circuit protective device.

**Q**

*Example 2*

A two-core, 16 mm² (6 mm² bare c.p.c.) PVC insulated and sheathed cable having copper conductors to BS 6004 (Ref 6242Y) is selected for a circuit. If $C_a = 30°C$, $Z_e = 0.35$ Ω and the cable's length of run is 40 m, determine if the circuit meets the requirements of Reg 543–01–03 of the *IEE Wiring Regulations*. Assume that the circuit is protected by a 60 A BS 1361 fuse and the wiring is installed using Reference Method 4 (Table 4D2A Wiring Regulations). Assume $U_0$ is declared at 240 V.

**A**

*Solution*

From Table 9A of the *IEE On-site Guide*, $(R_1 + R_2) = 4.23$ mΩ. Since Table 54C applies (1.38 factor) and the length is 40 m, then:

$$R_1 + R_2 = 4.23 \times 40 \times 1.38 \times 10^{-3}$$
$$= 0.233\ \Omega$$

The actual earth fault loop impedance of the circuit is:

$$Z_s = Z_e + R_1 + R_2$$
$$= 0.35 + 0.233 = 0.583\ \Omega$$

The fault current is found to be:

$$I_f = U_0/Z_s$$
$$= 240/0.583$$
$$= 412\ \text{A}$$

From the time/current characteristics in Appendix 3 of the *IEE Wiring Regulations*, the 60 A BS 1361 fuse will operate in 2 s. Using the formula in Reg 543–01–03, then:

$$S = \sqrt{I^2\ t/k}$$
$$= \sqrt{(412^2 \times 2)/115}$$
$$= 5.06\ \text{mm}^2$$

The calculated size of c.p.c. is smaller than the chosen c.p.c. and the requirements of Regulation 543–01–03 are met.

**Q**

*Example 3*

An XLPE armoured cable (solid aluminium conductors), complying with BS 5467 is chosen to feed a factory's 3-phase, distribution board taking a load of 50 kW/415 V at p.f. 0.7 lagging. The cable is 50 m long and installed in free air (installation Method 13). If the ambient temperature is 25°C and the circuit protective devices are BS 88 fuses, determine a suitable size cable for the installation. Assume the disconnection time and thermal constraints are both satisfied and $U$ is declared at 415 V.

**A**

*Solution*

Design current:

$$I_b = P/(\sqrt{3} \times V_L \times \text{p.f.})$$
$$= 50,000/(1.732 \times 415 \times 0.7)$$
$$= 99.4\ \text{A}$$

Hence $I_n$ is 100 A.

The only correction factor that applies is $C_a$ which is 1.2 (see Table 4C1, Wiring Regs) and therefore the tabulated current carrying capacity of the cable is found to be:

$$I_t \geq 100/1.2$$
$$\geq 83.33 \text{ A}$$

The voltage drop allowance for the cable is given by:

$$V = 4\% \times U$$
$$= 0.04 \times 415 = 16.6 \text{ V}$$

From Table 4L4A, columns 1 and 5, a 25 mm² cable carrying 98 A is selected. This cable has a voltage drop of 2.7 mV/A/m and using the formula to find the actual voltage drop:

$$V = L \times I_b \times \text{mV/A/m}$$
$$\text{then } V = 50 \times 99.4 \times 0.0027$$
$$= 13.42 \text{ V}$$

The cable can be used for the circuit.

**Note:** Figure 2.14 shows details of the cable used in the above example and you should note from the table that the 25 mm² aluminium cable has a conductor resistance of 1.2 Ω/km and an armouring resistance of 2.7 Ω/km. Any fault to earth will travel along these two conductors and if all joints provide negligible resistance and the phase conductor's reactance is ignored, the actual impedance of the 50 m route is 0.195 Ω. Applying Table 54B of the Wiring Regulations and using the factor of 1.48 as a multiplier, the

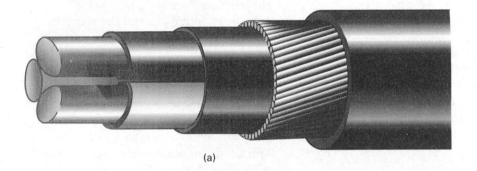
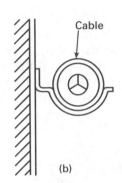

Cable

(a)

(b)

| Nominal area of conductor | Maximum resistance per km of cable at 20°C (ohms) | | | | | | | | | | | | | |
|---|---|---|---|---|---|---|---|---|---|---|---|---|---|---|
| | Copper conductor | Aluminium conductor | Armour resistance of copper conductor cables | | | | | | Armour resistance of aluminium conductor cables | | | | | |
| | | | Two core | | Three core | | Four core | | Two core | | Three core | | Four core | |
| mm² | | | PVC | XLPE | PVC | XLPE | PVC | XLPE | PVC | XLPE | PVC | XLPE | PVC | XLPE |
| 1·0 | 18·1 | – | – | – | – | – | – | – | – | – | – | – | – | – |
| 1·5 | 12·1 | – | 10·7 | 9·4 | 10·2 | 9·1 | 9·5 | 8·5 | – | – | – | – | – | – |
| 2·5 | 7·41 | – | 9·1 | 8·8 | 8·8 | 8·2 | 7·9 | 7·7 | – | – | – | – | – | – |
| 4 | 4·61 | – | 7·5 | 7·9 | 7·0 | 7·5 | 4·6 | 6·8 | – | – | – | – | – | – |
| 6 | 3·08 | – | 6·8 | 7·0 | 4·6 | 6·6 | 4·1 | 4·3 | – | – | – | – | – | – |
| 10 | 1·83 | – | 3·9 | 6·0 | 3·7 | 4·0 | 3·4 | 3·7 | – | – | – | – | – | – |
| 16 | 1·15 | 1·91 | 3·5 | 3·8 | 3·2 | 3·6 | 2·2 | 3·2 | 3·7 | 3·9 | 3·4 | 3·7 | 2·4 | 3·4 |
| 25 | 0·727 | 1·20 | 2·6 | 3·7 | 2·4 | 2·5 | 2·1 | 2·3 | 2·9 | 4·1 | 2·5 | 2·7 | 2·3 | 2·4 |
| 35 | 0·524 | 0·868 | 2·4 | 2·5 | 2·1 | 2·3 | 1·9 | 2·0 | 2·7 | 2·9 | 2·3 | 2·5 | 2·0 | 2·2 |
| 50 | 0·387 | 0·641 | 2·1 | 2·3 | 1·9 | 2·0 | 1·3 | 1·8 | 2·4 | 2·6 | 2·0 | 2·2 | 1·4 | |
| 70 | 0·268 | 0·443 | 1·9 | 2·0 | 1·4 | 1·8 | 1·2 | 1·2 | 2·1 | 2·3 | 1·4 | 1·9 | | |
| 95 | 0·193 | 0·320 | 1·3 | 1·4 | 1·2 | 1·3 | 0·98 | 1·1 | 1·5 | 1·6 | | | | |
| 120 | 0·153 | 0·253 | 1·2 | 1·3 | | | | | | | | | | |
| 150 | 0·124 | 0·206 | | | | | | | | | | | | |
| 185 | 0·0991 | | | | | | | | | | | | | |
| 240 | | | | | | | | | | | | | | |

(c)

Figure 2.14    Conductor and armour resistance values – 600/1000 V cables: (a) typical XLPE 3-core aluminium armoured cable; (b) cable cleat (installation method 13); (c) table containing resistance values of copper and aluminium cables

impedance increases to 0.289 Ω. This figure would then be added to the external earth loop impedance value for the type of earthing system used. This would then give the actual earth loop impedance of the circuit which is required for finding the fault current.

*Example 4*

Using Table 4J1A of the *IEE Wiring Regulations*, determine what size 2-core MI cable can be used for a 230 V single-phase heating load rated at 5.76 kW. The cable has a PVC sheath and is 25 m in length, clipped on a wall and in an ambient temperature of 40°C. A type B m.c.b. is used to protect the circuit. Both disconnection time and thermal constraints are satisfied.

*Solution*

$$I_b = P/U_0$$
$$= 5760/230 = 25 \text{ A}$$

$I_n$ is chosen to be 32 A. From Table 4C2 of the Wiring Regulations $C_a = 0.93$. Hence:

$$I_t \geq 25/0.93 = 26.9 \text{ A}$$

From Table 4J1A, cols. 1 and 2, a 2.5 mm² is chosen.

In the above example, if the MI cable was bare and exposed to touch, what affect would this have on the calculation and the selected size of cable? What would be the situation if the ambient temperature increased to 55°C?

---

## EXERCISE 2.2

1. With reference to the *IEE Wiring Regulations*, state the Regulation number and describe with the aid of diagrams the requirements concerning the following cables:
   (i) single-core armoured cables used in a.c. circuits;
   (ii) underground cables installed in a trench;
   (iii) cables installed overhead between buildings

2. Make a fully labelled sketch of a rising main busbar trunking showing the internal live conductors and internal fire barrier.

3. (a) What are the *IEE Wiring Regulations* requirements regarding BS 6004, PVC flat twin/c.p.c. cables run in floor joists?
   (b) Make a fully labelled sketch of methods that are used to protect the cables in (a) above.
   (c) In some circumstances, it is permissible to dispense with protective measures against indirect contact. What is the minimum length of concealed metal conduit that can be used without being earthed?

4. Make a fully labelled sketch of a typical suspension support bracket for lighting trunking.

5. (a) What are the Wiring Regulations regarding category circuits?
   (b) Make a sketch to show how different category circuits can be installed in trunking.

6. (a) State TWO types of wiring system suitable for each of the following electrical installations, giving reasons for your choice and quoting TWO requirements from the *IEE Wiring Regulations* that might apply:
   (i) boiler room
   (ii) petrol station forecourt
   (iii) domestic premises
   (iv) farm milking parlour.

7. PVC conduit and single-core PVC sheathed cables are chosen as the wiring system for a single-phase, 6 kW/240 V heating load. If overcurrent protection is by a BS 3036 fuse and the length of run is 30 m, select from the Wiring Regulations suitable size cables for the circuit. Assume the ambient temperature to be 35°C, the voltage drop limited to 4 V and both shock and thermal requirements met by installing a c.p.c. of the same size as the live conductors.

8. What are the Wiring Regulations' requirements concerning multicore armoured cables installed on cable tray with regard to the following?
   (i) Grouping
   (ii) Spacing of supports
   (iii) Bends in the cables

9. Using the *IEE Guidance Notes No.1 Selectio and Erection*, determine a suitable size 1 gauge BESA trunking for the following stran ed single-core copper cables:

thirty-eight 1.5 mm$^2$;
twenty-four 2.5 mm$^2$;
eighteen 6 mm$^2$;
nine 16 mm$^2$

**10.** State FIVE safety considerations for EACH of the following wiring systems when they are being erected.

(i) A number of armoured cables on cable tray/ladder.

(ii) A number of mineral insulated cables clipped together on a wall surface.

(iii) A rising main busbar system passing through floors.

# Inspecting, testing and commissioning work

---

## Objectives

After working through this chapter you should be able to:

1 state the requirements needed for a technical manual;
2 describe the procedure for the initial verification of an electrical installation;
3 state a number of common deviations from the requirements of the IEE Wiring Regulations;
4 know requirements for installation test equipment;
5 know the procedures for making periodic inspection and testing of electrical installations;
6 know a number of considerations in the design, inspection, testing and commissioning of:

(i)   emergency lighting systems;
(ii)  fire alarm systems;
(iii) petrol filling stations.

## Specifications and manuals

In this area of work, both the 16th Edition of the *IEE Wiring Regulations* and the *IEE Guidance Note No. 3, Inspection and Testing* should be consulted. The latter document requires the preparation of a full specification prior to the commencement or alteration of the electrical installation. It should set out the requirements for which the installation is to be used and include a description on how the installation is to operate, together with the necessary technical information and, where applicable, instructions of the installed electrical equipment. The specification could be provided by the designer, installer, owner or user of the electrical installation. It could also be provided by an authoritive body such as the HSE or a licensing authority or fire authority.

Of particular interest to those concerned in this area of work is BS 4884 which offers guidance on the preparation of technical manuals. Part of a typical operating manual to be given to the architect or client on hand-over of an installation would include the following.

### Operating and maintenance manual

- A full description of each system installed, written to ensure that the client's staff fully understand the scope and facilities provided.
- A full description of the mode of operation of all systems.
- Diagramatic drawings of each system indicating principal items of plant equipment.
- A photo-reduction of all record drawings to A4 size together with an index and legend for all colour coded services.
- Schedules of each system of plant and equipment stating their location, ratings and duty. Each item to have a unique number cross-referenced to the record, diagramatic drawings and schedules.
- The name and address and telephone number of the manufacturer of every item of plant and equipment together with catalogue list numbers.
- Manufacturers' technical literature for all items of plant and equipment, assembled specifically for the installation, including detailed drawings, electrical circuit details and operating maintenance instructions.
- A copy of all test certificates for the installation including electrical circuit tests, corrosion tests, type tests, start and commissioning tests.
- A copy of all manufacturers' guarantees and warranties.

- Starting up, operating and shutting down instructions for all electrical systems installed.
- Schedules of all fixed and variable equipment settings established during commissioning.
- Details of action to take in emergency conditions.
- Procedures for seasonal changeovers.
- Recommendations as to the preventative maintenance frequency and procedures to be adopted to ensure the most efficient operation of the systems installed.
- A list of normal consumable items.

## Verification

The first requirement to look at in terms of verification is Regulation 130–10–01 of the *IEE Wiring Regulations*. This being the last Regulation in Chapter 13 concerned with *Fundamental Requirements for Safety*. It partly states:

> On completion of an installation or an extension or alteration of an installation, appropriate tests and inspection shall be made, to verify so far as is reasonably practicable that the requirements of Regulations 130–01 to 130–09 have been met.

Figure 3.1 shows a diagram of the three main areas required by this Regulation. Chapter 71 of the *IEE Wiring Regulations* describes the procedure for the initial verification of an installation concentrating

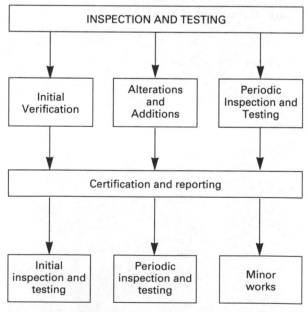

Figure 3.1  Process of inspecting and testing electrical installations

on general information, inspection and testing. Detailed inspection should precede testing with that part of the installation to be inspected, disconnected from the supply. Both inspection and testing methods are covered in Chapter 4 of the author's *Part I Studies: Theory*.

The *IEE Guidance Notes No. 3* refers to the person carrying out inspection and testing work to be competent, skilled, experienced and having sufficient knowledge of the installation to ensure that no danger occurs. This person (referred to as an 'inspector'), requires information about the installation's electrical design as described in Part 3 of the *IEE Wiring Regulations, Assessment of General Characteristics*. The installation's external earth fault loop impedance path ($Z_e$) and prospective short circuit current ($I_{pscc}$) are both required by Regulation 313–01–01 being determined by calculation, measurement, enquiry or inspection (see Note below).

**Note:** Both $I_{pscc}$ and $Z_e$ can be obtained from a regional electricity company (REC) as explained in Chapter 1 under Electricity Supply Regulations 1988 (see also Figure 1.2 showing the path taken by the earth fault current).

The Electricity Association produce a number of engineering recommendations, for example, P.23/1 which gives typical maximum values of $Z_e$ for different types of earthing arrangements or supplies up to 100 A. These are 0.35 Ω for a TN–C–S system, 0.8 Ω for a TN–S system and 21 Ω for a TT system. It should be noted that there is only one external earth loop impedance.

In large installations where a number of sub-mains are installed, marking distribution boards with $Z_s$ values is often carried out. This procedure may be omitted where the short-circuit capacity of every device within the installation is greater than the value of the $I_{pscc}$ at the origin of the installation.

The Electricity Association's P.25 gives values of estimated maximum $I_{pscc}$ at an RECs cut-out fuse based on a declared level of 16 kA at the point of connection of the service line into low voltage distribution mains. This declared maximum value is based on the information stated in Section 5 of the *IEE Guidance Note No. 6, Protection against Overcurrent*. For a TN–C–S service cable, the minimum value of $Z_s$ is given as 0.015 Ω and therefore the prospective short-circuit current between phase and neutral/earth at the tee-off point to the consumer's service line is found to be:

$$
\begin{aligned}
I_{pscc} &= U_0/z_s \\
&= 240/0.015 \\
&= 16 \text{ kA}
\end{aligned}
$$

It is important to note that an REC can install their supply transformer anywhere on the network. You should note that a 1 MVA, 11 kV/415 V supply transformer can develop an internal short circuit of approximately 27 kA. Fortunately, the lengths of main distributor cable and service line cable considerably reduce this value. You can see this in the following table of supply cables with conductor resistance dominating conductor reactance.

| Cable | Resistance (Ω/km) | Reactance (Ω/km) |
| --- | --- | --- |
| 300 mm² | 0.102 | 0.072 |
| 185 mm² | 0.165 | 0.074 |
| 120 mm² | 0.254 | 0.075 |
| 70 mm² | 0.444 | 0.075 |
| 35 mm² | 0.871 | 0.077 |
| 25 mm² | 1.200 | —— |

For domestic consumer installations, it is usual for the supply authorities to run a 25 mm² PVC concentric service cable to the cut-out position. This cable has a current rating of 150 A (Cu) or 115 A (Al) when run in the ground. The installation is protected by a 100 A, BS 1361 Type II cartridge fuse which can handle short-circuit currents up to 33 kA.

## Visual inspection

The inspector's responsibility in making visual inspections is to ensure that the correct materials (designed to recognised standards) have been properly selected and installed and that connected equipment is seen to be undamaged and appropriate for the environmental conditions. Listed below are some typical deviations from the *IEE Wiring Regulations* that might be found when making a visual inspection of a new installation.

- No compliance with good workmanship and proper materials to Reg. 130–01–01.
- Equipment not in compliance with British Standards or other recognised EC Standard.
- The installation not suitable for the environment (e.g. steel screws used outside and aluminium/brass used with copper conductors without protection).
- Equipment not selected with the correct IP reference.
- Ingress of water or dust in an enclosure owing to missing blanking plates.

- Equipment not suitable for a.c. voltage and current.
- Phase conductors of single-phase circuits not identified in red.
- Neutral conductor not connected in the same sequence as the phase conductor.
- Use of a 'borrowed' neutral.
- Residual current device, fed from a 13 A socket outlet, not rated at 30 mA for equipment used out of doors.
- Omission of the quarterly r.c.d. test notice.
- Omission of a label to indicate that a socket outlet is suitable for outside use.
- Omission of the main equipotential bonding conductor to the service water/gas pipes.
- Main bond connection not as close as practicable to the point of entry.
- Exposed conductive parts of the installation not earthed.
- Inadequate sizes of main and supplementary bonding conductors.
- Safety label missing on BS 951 earth/bonding clamp.
- Omission of supplementary bonding in bathroom/shower rooms.
- Omission of Home Office skirt on lampholder in bathroom.
- Quality of the circuit protective device not suitable for the prospective short-circuit current.
- Omission of the main switch or circuit breaker.
- Voltage drop not considered in the installation.
- Confusion between identification labels for switchgear and controlgear.
- Omission of the periodic inspection label – Reg. 514–12–01.
- No provision of charts, diagrams or tables.
- Appliances and luminaires not in readily accessible position.
- No mechanical protection for cables – see Regs 522–06.
- Holes in joists for cables not 50 mm from the top or bottom of joists – Reg. 522–06–05.
- Cables buried in walls not being horizontal or vertical – Reg 522–06–06.
- Omission of cable covers or marker tape for underground cables.
- Mixed category circuit wiring and omission of barriers in dual boxes for 13 A T.V. socket outlets.
- Wrong use of twin socket outlets on a ring final circuit supplying equipment having high earth leakage current.

- Poor joints and omission of cord grip devices.
- Undersize terminations used on cables and teminations used without lugs.
- Entry holes made too large at ceiling roses.
- Failure to make good holes in walls and ceilings.
- Insufficient clipping of cables, including the roof voids.
- Insufficient supports for conduits and trunking.
- Failure to issue completion and inspection certificate

## Testing and measurements

Section 713 of the *IEE Wiring Regulations* and Sections 7 to 16 of the *IEE Guidance Notes No. 3, Inspection and Testing* should be consulted. It is important to carry out testing with the installation and load equipment isolated and proved to be dead. Some tests such as earth fault loop impedance and residual current devices will of course require a supply. The following test instruments should be available.

- **Continuity (low resistance) tester** – (low scale resolution of 0.01 $\Omega$ open circuit voltage 3–24 V, test current > 20 mA).
- **Insulation (high resistance) tester** – (output current 1 mA at d.c. test voltage 250 V, 500 V or 1000 V). Tests to achieve 0.5 M$\Omega$ for 500 V test and 1 M$\Omega$ for 1000 V test.
- **Earth fault loop impedance tester** – (low scale resolution of 0.01 $\Omega$, test current 20–25 A and max. test duration 40 ms).
- **Residual current device tester** – (test currents 10–500 mA). Tests to achieve no operation at a tripping current of 50% for a duration of 2 s and one at 100% to operate the device in less than 200 ms. Test to achieve operation at tripping current of 150 mA to be less than 40 ms with maximum test time 50 ms.
- **Earth electrode tester** – (scale resolution 0.01–20 $\Omega$ at 10 mA or 0.1–200 $\Omega$ at 1 mA, frequency 128 Hz). Test results for TN and TT systems must not exceed 50 V.
- **Other test instruments** – voltage indicator, portable appliance tester, cable fault locator.

## Periodic inspection and testing

Chapter 73 of the *IEE Wiring Regulations* is concerned with periodic inspection and testing and is further described in Section 5 of the *IEE Guidance*

*Notes No. 3.* It is important to stress the safety of persons, livestock and property against the effects of shock, burns and damage when making tests, especially if there are defects on the installation. Regulation 731–01–02 (iv) of the Regulations requires any defects or non-compliance to be identified.

In making tests in some large, existing installations it is necessary, initially, to carry out a survey. The test inspector will require design data, charts, drawings, schedules, etc., and all relevant information about the installed circuits and the connected equipment. It is important to relate the findings from the periodic inspection and testing to the requirements of the Electricity at Work Regulations 1989.

When carrying out periodic inspection reports and schedules, the National Inspection Council for Electrical Installation Contracting (NICEIC) requires every periodic inspection report from its approved contractors to be accompanied by appropriate schedules (i.e. a schedule of the items that have been visually inspected, a schedule of the items that have been tested in support of the visual inspection, and a schedule of the results of the inspection and tests). The NICEIC emphasises the importance of agreeing with the client or third party about the extent and any limitations in carrying out the periodic inspection and testing work.

---

# EXERCISE 3.1

1. Make a list of the possible errors an electrical operative might make when wiring the following accessories:
   (i)   single-pole switches
   (ii)  Edison-type lampholders
   (iii) ceiling roses
   (iv)  switch sockets and spur boxes
   (v)   joint boxes

2. Describe with the aid of diagrams how you would make the following tests on an installation:
   (i)   ring circuit continuity
   (ii)  insulation resistance
   (iii) earth fault loop impedance
   (iv)  earth electrode resistance
   (v)   residual current device

3. State the electrical test equipment required by an electrical contractor.

## Certification

Chapter 74 of the *IEE Wiring Regulations* and the *IEE Guidance Note No. 3* are concerned with certification. The Notes contain a number of forms which should be read and adapted for the type of premise being inspected and tested. These forms are:

- WR1 – completion and inspection certificate
- WR2 – particulars of the installation
- WR3 – form of inspection
- WR4 – form of test
- WR5 – installation schedule
- WR6 – forms of periodic inspection and test for an electrical installation
- WR7 – minor works.

**Note:** The forms WR1 to WR5 are used when inspecting and testing new installations.

Other certificates and forms required to be used for special installations are:

- Emergency lighting certificate to *BS 5266 Part 1 1988, Emergency Lighting.* Appendix B is used for the initial completion of an emergency lighting installation or part thereof.
- Emergency lighting inspection and test certificate to the above BS 5266. Appendix C is used for the initial inspection and all subsequent periodic inspections and tests of existing installations (Clause 11.2).
- Fire alarm installation and commissioning certificate to *BS 5839 Part 1 1988, Fire detection and alarm systems for buildings.* Appendix B is used for all new installations following inspection (Clause 26.6).
- Fire alarm testing certificate to the above BS 5839. Appendix C is used annually and in conjunction with a 5-year inspection and test of the electrical installation (Clauses 29.2.7 and 29.2).
- Petrol filling station certificate of electrical inspection and testing to HSE publication, *HS(G)41, Petrol filling stations: construction and operation.* Appendix 4 covers initial completion or periodic inspection of an existing electrical installation.

---

## Commissioning work

This is the area concerned with the functional operation of every installed electrical system in a new premises. It should be carried out in stages in order to quickly identify any problems or faults that may occur. The person, a **commissioning engineer** from a specialist commissioning company, will instruct staff of the occupied premises on how to operate the installed equipment.

Commissioning work may take several days to complete, depending upon the complexity of the installation. In general, a subcontractor will issue the commissioning engineer with information concerned with the programme, specification, working drawings, control system diagrams, work test certificates and performance curves. The programme must indicate the stages for each system from pre-commissioning through to final handover and must also indicate the input and attendance required from all works contractors, manufacturers and suppliers of specialist equipment. It is important that a copy of the commissioning document is kept on site where it can be regularly upgraded as the commissioning work proceeds.

The commissioning engineer will prepare the commissioning document containing all checks and test results necessary for each individual system. Typical documents that he will require are:

- witness certificate
- system progress sheet
- pre-commissioning check sheet
- defects sheet
- fire dampers check sheet
- initial fan/pump test sheet
- distribution/outlet test sheet
- final fan/pump test sheet
- schematic diagrams
- plant thermal check sheet
- environmental check sheet
- control test sheet

In terms of a practical example, consider a gas heating system. The commissioning engineer will need to be familiar with the manufacturer's specification manual. This will provide him with a full description of the installation and the control sequence of operation. Once the boiler system has been filled with water and the gas supply switched on, the commissioning engineer can operate the gas burner, checking the flame failure mechanism and make necessary adjustments for correct gas pressure. He will also need to make gas tests. With the initial switching-on of the system, sensors, such as thermostats on the equipment and around the building need to be checked for their correct temperature settings. Checks should be made of the system to

see that heat transfer is not causing damage to other equipment. There should also be checks made on ventilation mechanisms and flow switches to see that they operate correctly and in compliance with the manufacturer's specification manual.

If the commissioning work involves electric motors, the engineer will need to make run-up tests of performance with observation made of each motor's operating control sequence (i.e. starting, running, timing, speed, etc.). The commissioning work should extend to checking foundations, shaft alignment, vibration, unusual noise, pulley belt tensions as well as control equipment, time lags, thermistors and other protective devices etc.

Typical clauses in a specification might read as follows.

### Additional inspection work

The specialist commissioning engineer to carry out the following.

With all control circuits disconnected, all isolators closed and all circuit protective devices fitted, the control panel should be subject to a voltage pressure test of 2 kV for 1 minute across the following points:

(i)  phase to phase
(ii)  phase to neutral
(iii)  phase to earth
(iv)  phase to neutral.

Two speed motors shall be tested to ensure that they can start their associated equipment at low speed.

Each main item of plant so tested shall be stamped or fitted with a label which shall give:

(i)  normal working conditions/duty
(ii)  test conditions
(iii)  equipment number
(iv)  date of test
(v)  name of manufacturer.

Each main item of plant shall be given a test certificate which shall include a witness signature.

Wiring terminations to all control equipment supplied shall be checked both for compliance with the wiring diagrams and interlocks as indicated in the specification.

The electrical subcontractor is responsible for the whole electrical installation which includes control panels, control wiring and all wiring systems.

## Emergency lighting

Emergency lighting is defined as lighting provided for use when the supply to normal lighting fails. There are a number of Acts, statutory instruments,

EC directives and codes of practice relating to this Category 3 system. Some of these are as follows:

- *The Cinematographic Act, 1952*
- *Fire Precaution Act, 1971*
- *The Health and Safety at Work etc. Act, 1974*
- *BS 5266 Part 1 1988* (see above)
- *BS 4533 Luminaire Product Standard*
- *BS 5499 Exit Signs Product Standard*
- Harmonised European standards (e.g. *CEN 169 WG3 Lighting Requirements, HD 384–5 (Wiring Standards)*

### Design requirements

The new European standard CEN 169 WG3 requires 1 lux on the centre line of an escape route (corridor, staircase, etc.) as the minimum light level allowed. BS 5266 requires 0.2 lux which is sufficient for permanently unobstructed escape routes. In open areas (offices and shops etc.), BS 5266 requires 1 lux and CEN 169 WG3 requires 0.5 lux. In high risk task areas (control areas, moving machinery etc.), CEN 169 WG3 requires illumination equivalent to 10% of the normal lighting to be provided within 0.25 s. If the European standard is implemented, testing of emergency lighting systems will become a legal requirement and tests will have to be carried out monthly and annually and records of the tests kept.

**Note:** Emergency lighting systems designed to meet the above requirements are classified as:

(i) self-contained;
(ii) small central systems;
(iii) large central battery.

Self-contained systems are luminaires containing their own built-in battery charging system and control gear. They operate either as a non-maintained system (emergency lamp off when the supply is healthy) or a maintained system (the emergency lamp having dual use with the mains supply). A sustained emergency luminaire is also used, having one lamp on when the supply is healthy and a different lamp on when the supply fails. In terms of emergency luminaires, BS 5366 requires exit signs to be consistent. In all new buildings exit signs should conform to BS 5499 (ISO6309) showing the international pictogram of a running man (see Figure 3.2).

The small central systems consist of compact units using sealed lead acid batteries supplying emergency power to a local area. The large central

Test switch

Figure 3.2 Emergency exit sign

battery system is a system with a sophisticated charging arrangement in which the batteries for a number of luminaires are housed in one location. The system may serve one or more final circuits or all emergency lighting luminaires in the building.

### Inspection and testing

| | |
|---|---|
| Daily | Check panel indicator lights and all maintained luminaires for correct operation. |
| Monthly | Test for a short period not exceeding one quarter of the duration. |
| Six monthly | Test as above for at least 1 hour for 3-hour rated systems. |
| Three-yearly | Test as above for full duration. |
| Subsequently | For self-contained systems, test annually for full duration. |

### Commissioning work

The commissioning engineer will inspect the installation to see that it conforms to the above regulatory requirements. This will include verification and operation of the luminaires by either temporarily removing the circuit protective device or operating a key switch facility. The engineer should also check the condition and state of batteries and their charging methods according to manufacturers' instructions.

## Fire Alarms

A fire alarm system is defined as a system of fixed apparatus for giving an audible/visible/perceptible alarm of fire. It may also initiate other action. Some of the principal regulatory requirements for a fire alarm system are as follows:

- *Fire Precautions Act, 1971*

- *Health and Safety at Work etc. Act, 1974*
- *Private Places of Entertainment Act, 1976*
- *BS 5839 Part 1, 1988* (see above)
- *BS 5445 Specification for components of automatic fire alarm systems*
- *BS 5588 Fire precautions in the design, construction and use of buildings.*

*Design requirements*

A fire alarm system is designed to protect life or property. There are three main classifications, namely:

(i)   Type L system to protect life;
(ii)  Type P system to protect property;
(iii) Type M system which is manually operated (see Note below).

Life and property systems have to be automatically activated but should also be equipped with manual call points. Type L systems can be divided into three sub-categories, namely L1, L2 and L3. Type L1 is when the fire alarm system is used throughout a building, type L2 is used only in defined parts of a building and type L3 is used only for escape routes. Type P is divided into type P1 and type P2. The former used throughout a building and latter used in defined parts.

For accurate identification of a fire, a fire alarm system is divided into **zones**. Each zone is normally restricted to one floor only – up to a maximum floor area of 2,000 square metres as in Figure 3.3. This can be relaxed where the total floor area of a building is less than 300 square metres.

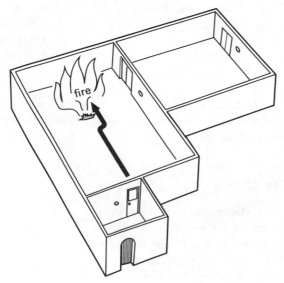

Figure 3.3   Diagram of a zone (single fire compartment)

The distance travelled by a person entering a zone to raise an alarm must not exceed 30 m. All alarm sounders should make the same sound. They should create a minimum sound level of either 65 dBA (decibels) or 5 dBA above any background noise likely to persist for a period longer than 30 s, whichever is the greater. For waking sleeping persons in hotels and boarding houses the minimum sound level at bedheads should be 75 dBA. At least one sounder per fire compartment is required and at least two sounders are required inside a building (arranged on at least two separate circuits). Reference should be made to the *IEE Wiring Regulations* requiring safety circuits to be independent of other circuits .

It is important to site the control panel in an area of low risk and preferably on the ground floor of a building by the entrance. An alarm sounder should be sited adjacent to the control unit. Standby supplies are normally from secondary batteries with automatic charging. They must have a life expectancy of four years. When the mains supply fails the standby batteries must be capable of operating the alarm system for 30 minutes after a certain minimum duration depending upon the building occupancy and the type of fire alarm system installed. For protection against life, if a mains failure is recognised within 12 hours, the battery system must operate for 24 hours.

The wiring system should preferably be MIMS cable to BS 6207 with or without PVC sheath. Other types of wiring system can be used provided their suitability can be clearly demonstrated. The cables should be routed in areas of low fire risk and protected against mechanical damage. Where installed in damp or underground locations they should have an overall PVC sheath and where installed in voids they should be separated from other cables by 300 mm – unless enclosed in metalwork such as conduit or trunking. The system should be regularly inspected and tested.

**Note:** Manual break glass call points (coloured red) must be sited in accessible places which are well lit, unobstructive and mounted at a height of 1.4 m. The call points should, where practicable, be of a common design and operate in a similar manner.

Where heat and smoke detectors are used, they should be sited so as to discriminate between fire and smoke in the normal environment. Heat detectors which respond to temperatures surrounding one particular location are called 'point type' heat detectors and those that respond to temperature changes along a line are called 'line type' heat detectors. The point type includes a fixed temperature element operating at a pre-determined temperature and are suitable for boiler rooms and kitchens. Some may have a rate-of-rise element which operates in response to a rapid rise in temperature and would be suitable for a drying room or enclosed loading bay.

Smoke detectors are either the ionisation chamber type or the optical scatter chamber type. In the ionisation type, a current flows between two electrodes which is reduced by smoke and when a predetermined value is reached the alarm circuit operates. They are sensitive to small-particle smoke from rapidly burning fires and find considerable use in offices, dining rooms, recording studios, etc. The optical type detector works on the principle of light scatter. When smoke enters the optical chamber the beam of light is scattered and reflection from the trapped smoke particles is detected by a photoelectric cell which initiates the alarm circuit. These detectors are more sensitive to larger particles of dense smoke, produced by slow burning fires and smouldering fumes. They are used in corridors and stairways where air currents may exist and also in guest rooms and hospital wards, etc. where smouldering fires may occur. For more details on their siting and spacing, you should consult manufacturers' catalogues on fire alarm systems.

*Inspection and testing*

| Daily | Check the control panel indicates normal operation. Check charger 'on' indicator. Check log book for outstanding faults. |
|---|---|
| Weekly | Operate a manual call point or detector to test the fire alarm system. Each week, operate a different detector. Check that all alarm sounders operate and reset the control panel. Check the standby battery connections. Check all call points and detectors, seeing that they are not obstructed. Enter result findings in log book. |
| Quarterly | Check entries in log book and take the necessary action. Examine the batteries and connections. Operate a manual point or detector to ensure that the system and all sounders operate. Check operation of the control panel. Enter results findings in log book. |
| Annually | Carry out checks and tests for the quarterly inspection. Test all call points and detectors. Check all cable fittings and equipment for damage. |

Every 2–3 yrs   Clean smoke detectors.

Every 5 years   Replace sealed lead acid batteries

*Commissioning work*

The commissioning engineer will inspect and test the installation to see that it conforms to the above regulatory requirements. The inspection will cover the manufacturer's recommendations in terms of siting call points and detectors, their suitability, specified height, number and spacing. The emergency battery system will be checked and tested for its proper working voltage and charging rate. The central/main fire alarm panel designed for the building will be tested, together with any zone indicator panel(s), repeater panel(s), magnetic door release unit(s) and any other ancillary equipment. Where the control panel incorporates connections for a remote alarm link via BT lines this also should be tested according to the manufacturer's instruction manual. After satisfactory completion of the tests, certificates in relation to Appendices A and B of BS 5839 Part 1, should be submitted to the architect/client/occupier of the building.

# Petrol filling stations

For these premises, some of the statutory and regulatory requirements that have to be complied with are as follows:

- *Petroleum (Regulations) Acts, 1928 and 1936*
- *Health and Safety at Work etc. Act, 1974*
- *Public Health Act, 1961*
- *The Safety Signs Regulations, 1980*
- *The Control of Substances Hazardous to Health, 1988*

- *The Electricity at Work Regulations*
- *BS 5345 Code of practice for selection, installation and maintenance of electrical apparatus for use in potentially explosive atmospheres, 1989* (numerous parts apply).
- *HSE, HS(G)41 Petrol filling stations: construction and operation, 1990.*
- *BS 1013 Earthing, 1965*
- *SFA 3002 Requirements for the certification of electrical systems in metering pumps for petrol filling stations (Buxton), 1971.*

## Design requirements

The planning and design of electrical installation work in a petrol filling station is covered in Part 3 of HS(G)41. This mentions that the facilities needed should be ascertained as accurately as possible by consultation between interested parties and organisations, such as the client, operator, architect, consultant, main contractor, electrical contractor, pump equipment manufacturer, fire insurer and licensing authority. The intention of such consultation is to prepare documentation about the installation and its related external conductive parts; accommodation and structural provisions for the filling station's equipment and its electrical services.

The guidance document suggests a register for retention of the documents at the petrol filling station. The electrical contractor would need to complete a record of the installation's general characteristics and schedules of all electrical circuits and equipment, and inspections and tests carried out. A summary of the other main areas covered in Part 3 of HS(G)41 are provided in the sections below. Some further explanation can be found on page 36, Chapter 3 of the author's *Electrical Installation Technology 3: Advanced*. Figure 3.4 also applies.

*Selection and installation of equipment*

This covers:

- equipment in hazardous and non-hazardous areas which requires electrical equipment to be constructed for the appropriate zones (see Note below);
- environmental conditions which requires electrical equipment to have a degree of protection against ingress of water or moisture and refers to *BS 5490 Index of Protection*.
- maintenance considerations which the installed electrical equipment is expected to receive during its intended life;

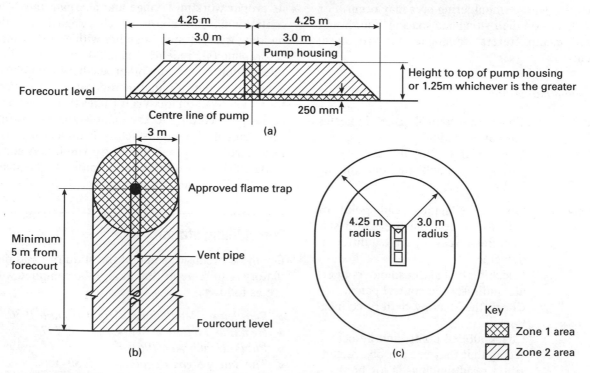

Figure 3.4  Classification of the danger areas at the forecourt of a petrol filling station: (a) pump housing; (b) vent pipe; (c) plan views from pumps

- lightning protection of buildings and structures to BS 6651 which requires the main earthing terminal of the electrical installation to be connected to the lightning conductor;
- radio and electrical interference to the limits prescribed in BS 800.

*Location of equipment*

This covers:

- dispensers for paraffin oil and diesel fuel meeting requirements for the appropriate hazardous area;
- battery charging equipment to be installed outside any hazardous area;
- vent pipes which should not be used to support luminaires or other electrical apparatus;
- canopies to be constructed above and clear of hazardous areas and any luminaires mounted underneath requires explosion protection;
- loudspeaker systems, preferably installed in non-hazardous areas or explosion-protected for the appropriate zone;
- luminaires are required to give an illuminance level of 100 lux over any dispensing area of the forecourt and protected where appropriate for its location;
- radio-frequency transmitting equipment installed such that it cannot induce a charge to a flammable atmosphere;
- socket outlets to be installed outside danger areas or appropriately protected for the environment;
- portable and transportable equipment to be appropriately protected and supplied at reduced voltage (110 V centre tapped to earth) or earth monitored r.c.d. arrangement (see HSE, PM 32).

*Isolation and switching*

This covers:

- the secure isolation of electrical equipment from every source of supply;
- main switch requirements, located in a non-hazardous area;
- pump motors, integral lighting and ancillary circuits which are not intrinsically safe to be provided with a means of isolation and where appropriate with warning notices where more than one supply source is supplied;
- central control point(s) which requires attended

self-service filling stations to have a means of disconnecting forecourt circuits;
- emergency switching to disconnect the supply to all metering pumps/dispensers and their integral lighting;
- high voltage or neon signs which should not be located in hazardous areas and should be installed to BS 559.

*Short circuit and overload protection*

This covers:

- circuits supplying dispensing equipments and requiring a single device to be used for short-circuit and overload protection with the device capable of dealing with the prospective short-circuit current at the origin of the installation;
- pump motors, integral lighting and ancillary circuits, requiring a suitable multiple-pole circuit breaker for isolation purposes.

*Protection against electric shock*

This covers:

- methods of protection against indirect contact by earthing and bonding or using Class II equipment;
- earthing systems but excludes a TN–C–S system from being used;
- main earthing bar or earth terminal position which should be accessible, located near to the point of supply and be provided with a test link;
- earthing of dispenser circuits in hazardous areas to be provided with a protective conductor;
- earthing bars or terminals in equipment enclosures except those having Class II construction;
- metallic conduit, ducting and trunking requiring a separate protective conductor of not less than 2.5 mm$^2$ when passing through a hazardous area;
- bonding of extraneous conductive parts which do not form part of the electrical installation;
- interconnection of the earthing systems (e.g. electrical installation, lightning installation, liquefied petroleum gas (LPG) installation or other installation) to ensure that all metalwork is at or about the same potential;
- continuity of bonding connections to ensure good electrical continuity with bonded metalwork having a resistance not greater than 0.01 Ω m at 20°C
- LPG installations requiring their earthing and bonding.

*Wiring systems*

This covers:

- the wiring installation within the hazardous areas but not in factory assembled units;
- conductor material which requires any conductor under 16 mm$^2$ to be made of copper, any aluminium protective conductor is not to be used;
- cables for intrinsically safe circuits which should be separated from other safe circuits;
- cables for extra low voltage circuits requiring the same insulation as high voltage circuits if contained in a common enclosure;
- cables installed underground requiring them to be at a depth of 0.5 m or be otherwise protected against mechanical damage; it also requires precautions to avoid the passage of flammable vapour from one hazardous zone to another by suitable sealing;
- protection against mechanical damage by moving vehicles;
- types of cables, preferring MIMS to be used in Zone 1 and 2 hazardous areas with armoured PVC sheathed cables also acceptable.

*Initial inspection and testing*

This requires the petrol filling station to be inspected and tested in accordance with the requirements of the *IEE Wiring Regulations* and in accordance with the requirements of the guidance notes and its relevant appendices.

*Periodic inspection and testing*

This requires an annual inspection and test programme to be carried out by a competent person to ensure that the electrical equipment and installation continues in a satisfactory condition. There is a need to use intrinsically safe test instruments where additions and alterations to the installation have been made and where it is not reasonably practicable to gas free the area. The programme should comprise:

- visual examination
- earth fault loop impedance testing at origin
- earth electrode testing
- continuity testing of protective conductors and equipotential bonding
- operation of residual current device
- insulation resistance testing
- recording and certification as indicated in Appendices 1 and 4.

**Note:** BS 5345 classifies hazardous areas as zones, namely:

Zone 0 – in which an explosive gas/air mixture is continually present or present for long periods;

Zone 1 – in which an explosive gas/air mixture is likely to occur in normal operation;

Zone 2 – in which an explosive gas/air mixture is not likely to occur in normal operation and, if it occurs, it will exist only for a short time.

## Commissioning work

It is pointed out in HS(G)41 that before a filling station can be brought into operation, it should be inspected by the licensee or other competent person to ensure that its condition is safe for public access and use. Commissioning work involves making tests on tanks, pipelines and fittings and checking to see if all emergency equipment is properly installed and in proper working order. The work also requires checking to see if warning notices and other information is in place. The commissioning engineer will need to study the electrical installation inspection and test schedules as detailed in HS(G)41 and make necessary checks himself to verify that the schedules and pre-commissioning tests have been carried out.

The NICEIC provide its approved electrical contractors with a petrol filling station certificate which contains a schedule of important items to be verified. Some of these are as follows.

*Initial inspection*

- type of earthing arrangement
- presence of test socket adjacent to the intake point
- security of fixings, glands, conduit, stoppers, etc.
- equipment suitable for hazardous zones
- correct lamp ratings
- presence/security of guards
- sealing of ducts between hazardous and non-hazardous zones
- correct function of relays and protective devices
- protection against corrosion, weather, vibration, etc.
- isolators capable of being locked in 'off' position
- protection against danger from overhead conductors
- loudspeaker system correctly installed and operational

*Periodic inspection*

- damage to apparatus or wiring
- security of fixings, glands, etc.
- condition of enclosures, gaskets and fastenings
- condition of compound or oil seals
- condition of pump motor bearing
- evidence of maintenance
- equipment clear of dirt and dust
- correct relay operation
- site documents and records kept in order

*Testing*

- integrity of earthing and bonding
- earth fault loop impedance at test socket outlet
- integrity of insulation
- operation of r.c.d. if fitted

---

# EXERCISE 3.2

1. Regulation 514–09–01 of the IEE Wiring Regulations requires a durable copy of circuit information, in the form of a schedule, to be fixed adjacent to a distribution fuseboard. Show how this might be done for the final circuits found in a domestic premises under the following headings:

   *Circuit / Protective device / No. of points / Wiring*

2. (a) Where a residual current device (r.c.d.) is installed in a premises, what are the Wiring Regulations' requirements regarding a notice for its regular testing?

   (b) Explain with the aid of a diagram the operation of the r.c.d. and state the periodic tests that have to be made.

3. (a) State the legal requirement for the maintenance of portable and transportable electrical equipment used in offices.

   (b) When making a visual inspection of office equipment, what factors need to be considered in the frequency of maintenance?

4. In Regulation 744–01–01 of the *IEE Wiring Regulations*, the result of periodic inspection and testing are to be recorded in a report and given to the person ordering the work. List a number of useful guidance notes for a person compiling the report.

5. (a) Draw a neatly labelled wiring diagram of a method that can be used to test emergency lighting luminaires.

   (b) Write brief comments on the following when making an inspection of an emergency lighting system:
   - positioning of signs and luminaires
   - wiring systems
   - power supplies
   - operation of central battery system
   - operation of engine driven generating plant
   - operation of self-contained luminaires and signs.

6. (a) In a fire alarm system, briefly comment on the siting and spacing of smoke and heat detectors.

   (b) What type of fire alarm detector would be suitable for the places listed below:
   (i) large department store
   (ii) photographic studio
   (iii) cargo holding area.

7. (a) Explain what is meant by the term 'hazardous area' in a petrol filling station.

   (b) List some of the requirements for a forecourt and petrol pumps.

8. (a) Draw a fully labelled circuit diagram of a high voltage neon sign.

   (b) State the purpose and requirements of a Fireman's switch.

9. (a) State the requirements for a live polarity tester.

   (b) Explain how you would carry out live polarity testing on a 415 V main switch, stating all necessary measures to avoid danger.

# Special installations or locations

## Objectives

After working through this chapter you should be able to:

1 *state the protective measures that are used to safeguard against electric shock occurring from direct and indirect contact;*
2 *describe some of the basic regulatory requirements when installing wiring systems and equipment in the following installations and locations:*
   - *bath tub and shower basins*
   - *swimming pools*
   - *hot air saunas*
   - *construction sites*
   - *agricultural/horticultural premises*
   - *restrictive conductive locations*
   - *high earth leakage current equipment*
   - *caravans and caravan sites*
   - *highway power supplies and street furniture;*
3 *perform a number of exercise tasks associated with the installations and locations listed above.*

## Protection for safety

The special installations and locations listed in Part 6 of the *IEE Wiring Regulations* concentrate on an important topic, that of safeguarding against electric shock. The damp and wet conditions encountered in these particular locations are likely to lower a person's body resistance, making it more vulnerable to electric shock. You will see from the touch voltage curve in Figure 4.1 that a person in contact with 240 V must be released from this danger in a time of 40 ms to avoid any harmful effects.

It is generally recognised that electric shock is caused either by touching a conductive part that is normally made live (**direct contact**) or by touching an exposed conductive part made live by a fault (**indirect contact**). The measures of protection used to safeguard against these two conditions are shown in Figure 4.2. **SELV** is an extra low voltage system which does not exceed 50 V a.c. or 120 V ripple free d.c. It has to be electrically separated from other systems and requires no earthing.

Typical supplies for SELV are shown in Figure 4.3. Briefly, the motor generator must have windings that provide electrical separation equivalent to a BS 3535 safety isolating transformer. When an electronic device is used, it must not be capable of producing an internal fault that causes the output voltage to exceed the maximum for extra low voltage. If SELV does not exceed 25 V a.c. (60 V ripple-free d.c.) then no measures are needed to protect against direct contact. However, if the voltage does exceed the values mentioned then either **barriers** or **enclosures** to IP2X must be used. Alternatively, **insulation** must be capable of withstanding a test voltage of 500 V for 60 s. In restrictive conductive locations this is a requirement despite the SELV voltage used.

The measure to limit the discharge of energy can only be used for individual items of current-using

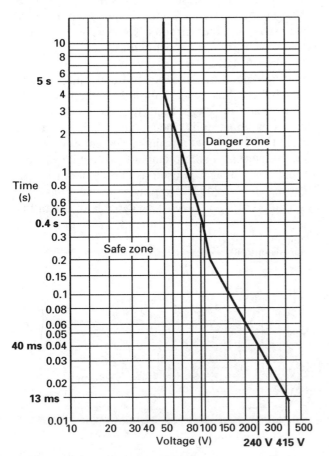

Figure 4.1   Touch voltage curve

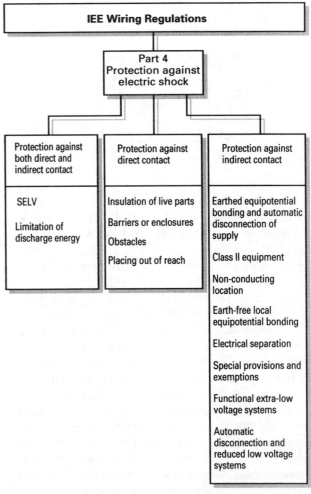

Figure 4.2   Measures of protection against electric shock

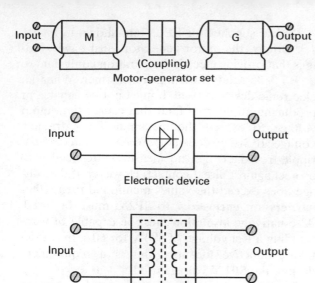

Motor-generator set

Electronic device

Isolating transformer

Battery

Figure 4.3   SELV supply systems

equipment which have been designed to a British Standard (or recognised EC standard). Like SELV, this method must be separated from any other circuit. An example is an electric fence controller (see Figure 4.4) which should not produce an energy pulse greater than 5 joules.

In terms of the measures used to protect against direct contact, insulation of live parts is a statutory requirement of the *Electricity at Work Regulations*. It is essential that insulation is capable of withstanding any mechanical, electrical, thermal or chemical stresses that may occur. Barriers and enclosures must be properly installed and maintained during the normal operation of the installation. Openings in enclosures which give access to live parts must be to IP2X (see Figure 4.5) but where this is unavoidable, such as replacing a lamp inside a lampholder, then suitable precautions must be taken to prevent live parts being touched. For top horizontal surfaces of barriers and enclosures which

are readily accessible, the degree of protection must be to IP4X. With the exception of ceiling roses, pull cord switches and lampholders, removing a barrier or opening an enclosure is only allowed with a tool such as a key or screwdriver. However, access is allowed if the supply is disconnected but its restoration should only be possible when all barriers or enclosures are in place.

Obstacles are used to prevent persons touching conductive live parts accidentally. They do not need to be removable with a tool but they have to be secured to prevent unintentional removal. The application of this measure is restricted to skilled or instructed persons and in some installations and locations it cannot be used.

Placing out of reach has similar requirements as obstacles and is used in locations accessible only to skilled persons or instructed persons. Overhead lines between buildings must meet the requirements of the *Electricity Supply Regulations*. Where bare live parts are out of arm's reach (see definition given in the Wiring Regulations) but are still accessible, they must not be within 2.5 m of exposed or extraneous conductive parts or bare live parts of other circuits. Considerable danger exists in farm and horticultural locations and Regulation 412–05–04 of the Wiring Regulations requires the above distance to be extended where long bulky objects are handled. Where two protective measures are used together, one of the measures must fully meet the requirements. It should be noted that a **residual current device** cannot be used as a sole means of protection against direct contact. If it is used as supplementary protection, its residual tripping current must not exceed 30 mA and it must be capable of operating

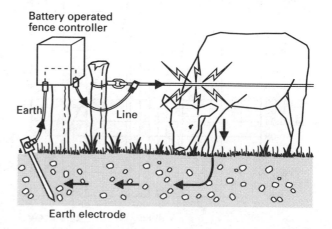

Figure 4.4   Electric fence controller

| FIRST NUMBER Degree of protection against solid objects | | SECOND NUMBER Degree of protection against water | |
|---|---|---|---|
| **0** | Non-protected. | **0** | Non-protected. |
| **1** | Protected against a solid object greater than 50 mm, such as a hand. | **1** | Protected against water dripping vertically, such as condensation. |
| **2** | Protected against a solid object greater than 12 mm, such as a finger. | **2** | Protected against dripping water when tilted up to 15°. |
| **3** | Protected against a solid object greater than 2.5 mm, such as a tool or wire. | **3** | Protected against water spraying at an angle of up to 60°. |
| **4** | Protected against a solid object greater than 1.0 mm, such as thin wire or strips. | **4** | Protected against water splashing from any direction. |
| **5** | Dust protected. Prevents ingress of dust sufficient to cause harm. | **5** | Protected against jets of water from any direction. |
| **6** | Dust tight. No dust ingress. | **6** | Protected against heavy seas or powerful jets of water. Prevents ingress sufficient to cause harm. |
| | | **7** | Protected against harmful ingress of water when immersed between a depth of 150 mm to 1 m. |
| | | **8** | Protected against submersion. Suitable for continuous immersion in water. |

Figure 4.5   Index of protection to IEC 529

in a time of 40 ms at a test current of 150 mA. See Figure 4.6.

The most common measure used for the protection against indirect contact is **earthed equipotential bonding** and automatic disconnection of the supply (see Figure 4.7). This measure is stated in Regulation 413–02 of the Wiring Regulations and is typically and partially provided by the components inside a distribution fuseboard or consumer unit. It simply means that if an electrical installation is properly designed and offers a low impedance path through its earthing, circuit protective devices and/or earth leakage protective devices should, as a result of an earth fault, operate automatically and

67

Figure 4.6 Typical residual current device used in conjunction with mains outlet socket

make circuits safe. The requirements for circuit protective devices to operate are based on **disconnection times** of 0.4 s for final circuits supplying socket outlets and 5 s for final circuits supplying only fixed appliances. Where automatic protection cannot be fulfilled by this method, then residual current devices may be used (see Note below).

Class II equipment relies on the provision of **double or reinforced insulation**. Regulations 413–03 and 471–09 of the *IEE Wiring Regulations* refer to the requirements of this measure. Although current-using equipment need no earth connection, a separate circuit protective conductor must be run to and terminated at each point in the wiring.

The measures known as **non-conducting locations** and **earth-free equipotential bonding** are special measures which are used under effective supervision and will not be discussed.

Protection by **electrical separation** is a measure that may be applied to single items of equipment such as a BS 3052 shaver unit or diagnostic testing of electronic equipment. The degree of electrical

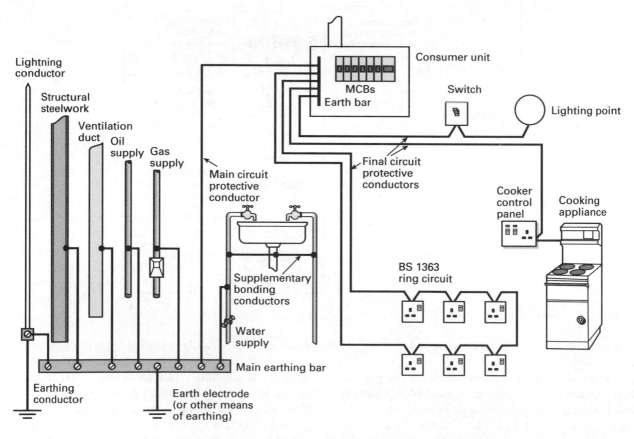

Figure 4.7 Earthed equipotential bonding and automatic disconnection of the supply

separation must not be less than that between the input and output of a safety isolating transformer complying with BS 3535 which may also be its source of supply. There must be no electrical connection between the separated windings and earth so that no path exists for current to flow to earth. Also, **overcurrent protection** is required in each pole of the separated circuit. When it is used to supply several items of equipment, in special situations, it should be under the control of a suitably qualified electrical engineer.

Special provisions and exemptions stated in Regulation 471–03 of the Wiring Regulations place emphasis on avoiding hazards in confined areas where skilled or instructed persons work on or near live equipment such as open-type switchboards. The requirements are for adequate space to safely operate and maintain equipment and requires suitable warning signs in areas where work is being carried out. Indirect contact need not be provided in some wiring situations which are isolated and inaccessible, such as lengths of metal conduit which do not exceeding 150 mm, fixing screws for non-metallic accessories and unearthed street furniture supplied from overhead lines. For further information, see Regulation 471–13–04 of the Wiring Regulations.

**Functional extra-low voltage** (FELV) is used where it is necessary to meet the requirements of the installed equipment, where not all SELV requirements can be satisfied. It is applicable where equipment has exposed conductive parts and requires earthing. A method called PELV is also used.

Where the use of extra-low voltage is impracticable, a reduced low voltage system may be used. This must not exceed 110 V a.c. between phases or 63.5 V a.c. between phase and earthed neutral. If the supply is taken from a centre-tapped double-wound isolating transformer the voltage will be 55 V.

If this measure of protection is used, overcurrent devices are required in each unearthed line conductor. The earth fault loop impedance at every point of utilisation, including socket outlets, shall be such that the disconnection time does not exceed 5 s. In these low voltage systems, every plug, socket outlet and cable coupler must have a protective conductor contact and their design must not be compatible with similar electrical accessories of other voltages or frequencies.

**Note:** In ordinary installations, the product of the circuit impedance (or resistance depending on the earthing system) and tripping current (or fault current) of a residual current device must not exceed 50 V. In some of the special locations dealt with in this chapter the voltage is reduced to 25 V. Socket outlets which can be reasonably expected to supply portable equipment for use outdoors of current rating 32 A or less must be provided with a residual current device as a means of supplementary protection against electric shock.

## Locations containing a bath or shower basin

The first requirement in Section 601 of the *IEE Wiring Regulations* is for protection against electric shock. This states that there should be no provision for electrical equipment to be installed in the interior of a bath tub or shower basin. For protection against both direct and indirect contact, SELV can be used provided the safety source is installed out of reach of a person using the bath tub or shower. For example, a transformer supplying a 12 V SELV extractor fan can be mounted in the loft area and the fan controlled either manually by a pull cord switch or automatically by means of a humidity sensor. Consideration should be given to the positioning of SELV circuits to avoid accidental splashing. Some fans used in these locations are designed to IP57.

The measures known as protection by non-conducting location and earth-free local equipotential bonding are both forbidden to be used. Except for SELV the earthing and bonding arrangements must be such that all circuit protective devices controlling final circuits must operate within a time of 0.4 s in the event of a fault. In view of the instantaneous tripping time of **miniature circuit breakers** that operate in 0.1 s, these circuit protective devices are highly recommended. Where supplementary bonding conductors connect two extraneous-conductive parts their sizes are limited to not less than 2.5 mm$^2$ if they are sheathed or provided with mechanical protection, or 4 mm$^2$ if mechanical protection is not provided.

Local equipotential bonding of the cold water and hot water pipes can be carried out in an adjacent airing cupboard without the need of a bonding cable from the main distribution board (see Figure 4.8). Metallic surface wiring systems and exposed earthing and bonding conductors are not to be used in these locations.

Mains BS 1363 socket outlets are not allowed in

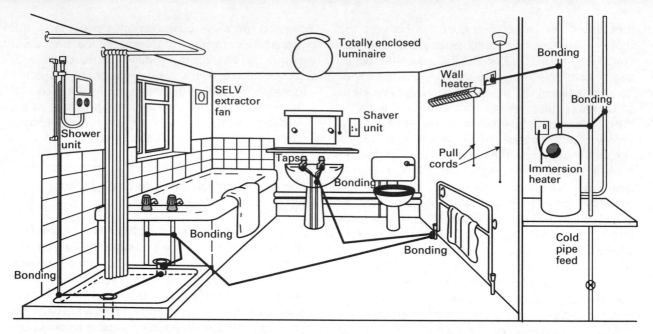

Figure 4.8  Electrical supplies and supplementary bonding in a bathroom and airing cupboard

a bathroom and there should be no provision for connecting portable equipment, except that a socket outlet shaver unit to BS 3535 is allowed. Where a shower cubicle is located in a room other than a bathroom, it should be installed at least 2.5 m from the shower cubicle. Consideration should be given to the distance of a 13 A socket outlet on a landing where it could be used, without thought, for the connection of a portable appliance inside a bathroom. Where a socket outlet is supplied from a SELV source the nominal voltage should not exceed 12 V and it must have no accessible metallic parts or be of the same pattern as BS 1363 socket outlets. The SELV circuit must not be included in the supplementary bonding and the 0.4 s disconnection time is not necessary.

A lampholder installed within a distance of 2.5 m from a bath or shower cubicle must be constructed of, or shrouded in insulating material. A BC lampholder (type B22) must be fitted with a protective shield (Home Office skirt) to BS 5042. Alternatively, a totally enclosed luminaire may be used.

Switches and other electrical controls should be situated as to be normally inaccessible to a person using a fixed bath or shower. Insulated cords and cord-operated switches, controls incorporated in instantaneous water heaters and mechanical actuators that are remotely controlled are all allowed in

these locations – provided they are designed to an appropriate British Standard or recognised foreign standard.

In terms of other equipment, an extractor fan, fixed electric wall-mounted heater and instantaneous water heater can all be installed provided they meet the above requirements. An electrical shower unit must be connected to the supply via a double-pole linked switch with a minimum contact gap of 3 mm in both poles. The switch must be in an accessible position and clearly identifiable and out of reach of the person using the shower – unless it is cord operated as mentioned above (see Figure 2.13).

If electric heating is embedded in the floor it must be covered by an earthed metallic grid or earthed metallic sheath and be bonded. Where a Jacuzzi or other pumping device is installed under a bath, its access must be by use of a tool. Where an electric towel rail is fed from an adjacent flex outlet the c.p.c. within the flex may be used as a supplementary bonding conductor. If a gas boiler is installed in a bathroom its controls should be inaccessible to a person using the bath or shower.

## Swimming pools

A SELV supply in these locations offers an ideal measure of protection against electric shock from direct contact. When it is used, whatever the nominal voltage, additional protection has to be provided by barriers or enclosures affording at least IP2X or IPXXB or insulation that must be capable of withstanding a test voltage of 500 V for 60 s. Other direct contact protective measures such as non-conducting location, placing out of reach, obstacles and earth-free equipotential bonding are not allowed. You will see from Section 602 of the *IEE Wiring Regulations* and Figure 4.9 that the area around the swimming pool is divided into zones. Zone A covers the pool area itself and any electrical equipment installed in this area such as pool lights, must be suitable for IPX8 (continuous immersion in water). The supply to each pool light must be from its own transformer (or an individual winding on a multi-secondary winding transformer) having an open circuit voltage not exceeding 18 V. Supplies to any other equipment must be from a SELV source situated outside Zones A, B and C

and have a nominal voltage not exceeding 12 V. The wiring should be of Class II construction without any metallic covering.

Zone B is an area extending from the pool edge for a distance of 2 m. Any electrical equipment installed in this area must be suitably protected to IPX4 (water splashing from any direction). Where water jets are used for cleaning the pool then the protection must be to IPX5. Only BS EN 60309–2 socket outlets can be installed in this zone, mounted at a height of 0.3 m from the floor and at least 1.25 m from the edge of the pool. The socket outlets have to be controlled by a 30 mA residual current device meeting the requirements of BS 4293. Water heaters complying with BS 3456 are permitted in this zone. Other equipment must be supplied from a 12 V SELV circuit.

Zone C extends a further 1.5 m beyond Zone B and any electrical equipment used in this zone must be protected to IPX2 for indoor pools and IPX4 for outdoor pools. In this zone, it is possible to install a shaver socket to BS 3535 and also a socket outlet, switch or accessory provided they are protected individually by electrical separation, SELV (not exceeding 50 V) or a residual current device to BS 4293. An instantaneous water heater to BS 3456 is not allowed.

In Zones A and B, no junction boxes are to be used and only the necessary wiring to supply appliances in these two zones is allowed. Supplies by SELV not exceeding 12 V can be used but the source must be situated outside Zones A, B and C.

In Zones B and C, heating units embedded in the floor must have an earthed metallic sheath connected to the local supplementary bonding conductors. They shall be covered by a metallic grid connected to the same local supplementary bonding conductors. Central heating radiators also require supplementary bonding and it is important to note that such bonding must not be made to appliances which are supplied from a SELV source. Where there are no exposed conductive parts in any of the above zones, the minimum size supplementary bonding conductor is 2.5 mm$^2$ if it is protected or 4.0 mm$^2$ if it is not protected.

In terms of main switchgear, boilers, pumps, etc., these are normally situated outside Zones A, B and C and the requirements of the Wiring Regulations must be met. The equipment should be suitable for swimming pools as well as the environment, particularly luminaires installed over the pool which may be affected by corrosion.

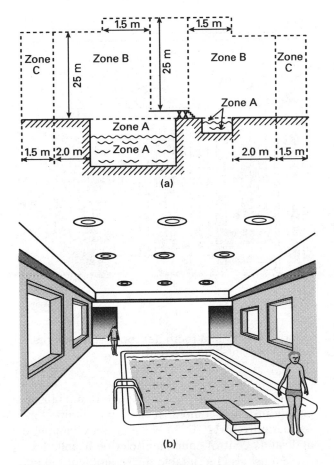

Figure 4.9 Swimming pool: (a) zone requirements; (b) internal arrangements

## Hot air saunas

A room built for a sauna bath is often constructed of wood panels or logs. Panel design is more likely to be used for an indoor sauna and consists of Finnish Northern Pine or Spruce. The walls and ceilings are made with an internal cladding of shiplap, porous insulation material and an external cladding of tongued and grooved timber. Log saunas have a similar design but the walls are solid baulks. The room should be provided with high-level outlet vents and low-level inlet vents that are internally controlled by a sliding wooden shutter. The entrance door is constructed in thermally insulated timber and has a spring-loaded ball catch.

The stove used for heating the room should be of stainless steel or stove-enamel steel construction and it should have a timber guard rail fitted to protect bathers from accidental contact. The stove is fitted with a thermometer and if supplied by electricity, its elements should be totally enclosed with the terminals enclosed in a waterproof connection box.

Electric elements vary in rating according to the internal cubic volume of the sauna room (e.g. a 4 kW element is adequate for a room size of 80–125 cubic feet). The stove should be capable of producing a room temperature of 100°C within a heat-up period of one hour. Stoves over 5 kW rating are normally operated through a contactor via a thermostat. The contactor's coil should be protected by a fuse and the complete stove circuit protected by a switch-fuse isolator.

The requirements for these locations are covered in Section 603 of the *IEE Wiring Regulations*. Protection against electric shock from direct contact is achieved by SELV and regardless of the nominal voltage must have insulation capable of withstanding a test voltage of 500 V a.c. for 60 s or barriers and enclosures providing protection to at least IP24 or IPX4B. Neither protection by means of obstacles nor protection by placing out of reach, may be used as a measure for direct contact. Similarly, protection from a non-conducting location or a means of earth-free local equipotential bonding, cannot be used as a measure for indirect contact.

Figure 4.10 shows the permitted zones of ambient temperature. All equipment must have at least IP24 protection. In Zone A, only the sauna heater and its associated equipment shall be used. There

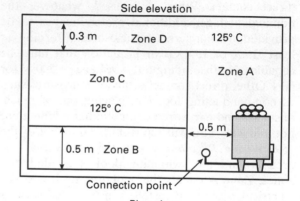

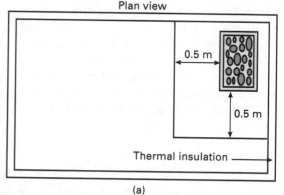

(a)

(b)

Figure 4.10   Sauna room: (a) zone requirements; (b) internal finishing of wood

are no special requirements in Zone B and in Zone C: any installed equipment must be suitable for a temperature of 125°C. In Zone D, only luminaires and sauna control equipment can be installed and these must also be suitable for an ambient temperature of 125°C.

72

The Wiring Regulations only permit the use of flexible cords with 150°C rubber insulation and these should be double insulated. Luminaires must be mounted to avoid overheating and the control switch is preferred to be sited outside and adjacent to the door of the sauna room.

## Construction site installations

These installations which include new building construction, repair, alterations or demolition of existing buildings and other work are dealt with in Section 604 of the *IEE Wiring Regulations*. Whilst supplies of electricity on construction sites are generally of a temporary nature, the installations themselves need to comply with the Wiring Regulations and other regulatory requirements.

Particular attention should be given to Regulation 341–1 of the Wiring Regulations calling for consultation with the person or body who will be responsible for the operation and maintenance of the installation. It is important that the protective measures for safety remain effective during the intended life of the installation and that regular period inspection, maintenance and repair are carried out. Temporary installations are required to be inspected and tested every three months.

In terms of safety, Regulation 604–02–02 in Section 604 of the Wiring Regulations requires different types of construction site equipment to be supplied at certain voltages. For example, SELV portable hand lamps used in confined and damp locations should not be operated above 25 V, portable tools and local lighting up to 2 kW can only be used on 110 V, fixed floodlighting can be used at 230 V and fixed or movable equipment rated above 3.75 kW can be given a supply of 400 V. The requirements for site offices, canteens, and other buildings must meet all the requirements of the *IEE Wiring Regulations*.

Generators used to provide temporary supplies on new sites should be earthed with the frame and neutral bonded. Their parallel operation with the public supply should not be possible unless prior agreement has been obtained from the regional electricity company. It is recommended that a generator is used in conjunction with a residual current device with its trip operation complying with the voltage in Regulation 413–02–16 reduced to 25 V.

Where small generators satisfy the need to supply power in areas where no site distribution has been installed and the arrangement is through an IT system, permanent earth fault monitoring is to be provided. Regulations 413–02–21 to 413–02–25 of the IEE Wiring Regulations have to be satisfied.

In three-phase systems the maximum disconnection times for circuits operating at different voltages, with or without neutrals are also reduced (see Table 604A in the Wiring Regs).

On a site provided with a TN–S earthing, reduced disconnection times are required for circuits supplying movable installations or equipment, either directly or through socket outlets.

In general, it is the preferred practice to use modular site distribution equipment to BS 4363 and BS EN 60439–4 (see Figure 4.11).

All switchgear should be installed in a position which allows free uninterrupted access during the course of site construction and it should be protected against damage and any adverse environmental conditions. The supply should be taken from an isolating switch which is capable of being locked in the off position. Every socket outlet must comply with BS EN 60309–2 and be incorporated in a BS 4363 and BS 5486 Part 4 equipment unit.

Cable wiring systems must comply with the Wiring Regulations and be properly sized and adequately protected, particularly when installed across site roads and walkways. There should be no deliberate strain imposed on cables or the terminations of conductors. Attention should be given to overhead supplies which run across sites and also cables which are run underground.

## EXERCISE 4.1

1. Figure 4.12 shows how a person can be protected from electric shock from indirect contact using the method called earthed equipotential bonding and automatic disconnection of the supply.
   (i) Assume the supply system is TN–C–S earthing, trace (with arrows) the route taken by the fault current and explain what happens during a fault to earth on the appliance.
   (ii) What is meant by fuse discrimination? If $F_1$ was a 100 A BS 88 Type 2 cartridge fuse, $F_2$ a 30 A Type 1 m.c.b. and $F_3$ a 13 A BS 1362 cartridge fuse, how would discrimination be achieved?
   (iii) What is the maximum disconnection time

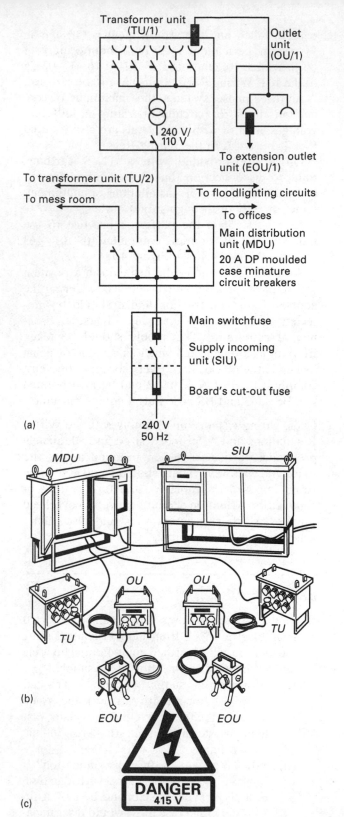

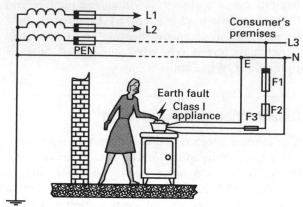

Figure 4.12 Protection against electric shock

for the appliance fuse, assuming it is connected through a 13 A socket outlet?

2. Figure 4.13 shows protection against electric shock by placing out of reach. What measure of protection is this and what are the regulatory requirements concerning the height of supply lines?

3. (a) What are the requirements of the *IEE Wiring Regulations* for supplementary bonding an electric towel rail in a bathroom?

   (b) What are the requirements of the *IEE Wiring Regulations* for installing socket outlets around an indoor and outdoor swimming pool?

   (c) On a construction site what type of plug and socket should be used for the connection of fixed flood lights and what is the maximum disconnection time for such circuits under earth fault conditions?

Figure 4.11 Construction site distribution system: (a) circuit diagram; (b) BS 5463 units; (c) warning sign to be marked on supply and transformer units. Note: SIU, MDU and TU are marked 'Danger Electricity'; OU and EOU are marked 'Caution Electricity'.

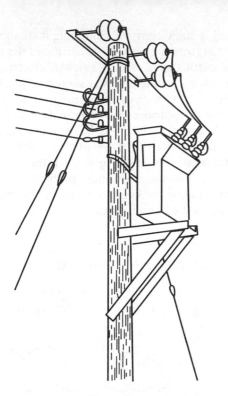

Figure 4.13    Protection by placing out of reach

## Agricultural and horticultural premises

The requirements for these installations can be found in Section 605 of the *IEE Wiring Regulations*. Farmhouse dwellings and other dwellings intended for human habitation are covered by other regulation requirements. Standard a.c. supplies to these installations are either single-phase, 240 V, 460 V (230 V – 0 – 230 V) or three-phase 400 V. Each building or group of buildings should be treated as a separate installation.

Protection for safety against electric shock is not only applicable to persons but also to livestock since they are quite easily electrocuted by a voltage of 25 V. The protective measures against both direct and indirect contact can be provided by SELV provided that the upper limit of 50 V is reduced for livestock. For protection solely against direct contact, barriers or enclosures to IP2X or IPXXB and/or insulation capable of withstanding a test voltage of 500 V for 60 s must be provided. Every socket outlet circuit must be protected by a residual current device. Protection solely against indirect contact is by earthed equipotential bonding and automatic disconnection of the supply.

For a TN earthing system the maximum disconnection times for circuits operating at 240 V is 0.2 s.

As mentioned before, mcb's provide ideal circuit protection but where fuses are used, lower earth fault loop impedance values are required (see Tables 605B1 and 605B2 of the Wiring Regulations).

The Regulations require supplementary equipotential bonding of all exposed conductive parts and extraneous conductive parts in locations which can be touched by livestock. Where a metal grid is laid in the floor it has to be connected to the protective conductors of the installation. This arrangement is often required in a milking parlour. If the earthing system is TT then the arrangement is carried out via an earth electrode and a residual current device.

Besides protection against electric shock, protection against fire and harmful thermal effects is also required. Here, a residual current device is required to be used for the supply to equipment and must not have a rating exceeding 500 mA (see Note below).

It is important that wiring systems are not located in areas where heat can build up by fortuitous insulation from bulk storage such as in stores housing cattle feed and food preparation, drying sheds, glasshouses and in buildings of light construction such as wooden poultry houses. In locations where livestock is kept, fixed wiring must be made inaccessible and where cables are liable to be attacked by vermin they must be of a suitable type and be suitably protected.

Supplies taken from one building to another should, preferably, be underground and the wiring system laid in pipes or ducts. Armoured cables laid directly in the ground should be adequately protected and identified with marker tape. Where overhead wiring is used, it must comply with the *IEE Wiring Regulations* and be adequately supported (see page 99, Table 13 of the *IEE Guidance Notes, No. 1*. Any emergency switching should be installed where it is inaccessible to livestock.

Electrical equipment which is accessible to livestock must be of Class II construction or suitably insulated. Radiant heaters must be kept at least 0.5 m from livestock and combustible material. Where quartz linear lamp heaters and metal sheathed infrared heaters are installed in milking parlours and dairies, etc., care should be taken to avoid water splashing the heaters. They should be switched off before any cleaning work takes place.

Fixed apparatus installed in a building where petrol-driven vehicles are used, stored or repaired should be mounted at a height above floor level determined by the local licensing authority or otherwise situated outside the area of risk. Equipment must be designed for the external influence encountered. Where portable and transportable equipment is used, cable couplers to BS 4343 and IP44 (minimum) should be selected.

Mains-operated fence controllers are required to comply with BS EN 61011 and BS EN 61011–1. Fences and controllers should be so installed and operated that they cause no danger to persons, livestock or surroundings. They should not be installed in places where there is a risk of fire. Only one controller is allowed to operate one fence and it should be connected to the supply through as double-pole switch, mounted in an accessible and visible position. It should not be fixed to any supporting pole associated with an overhead power line or telephone line. If the system earth of a fence is installed in the vicinity of a building it should be at least 10 m from the protective earth and the supply system earth (see Figure 4.14). An electric fence erected near a public highway should be fitted with suitable warning notices at frequent intervals.

**Note:** For the purpose of discrimination with other residual current devices, it is recommended that the 500 mA device is fitted with a time delay. Some consideration should be given to the problems caused by nuisance tripping, particularly where rcd's are used in milking parlours and in the heating of livestock houses, etc.

## Restrictive conductor locations

These locations are covered in Section 606 of the *IEE Wiring Regulations* and are applicable to

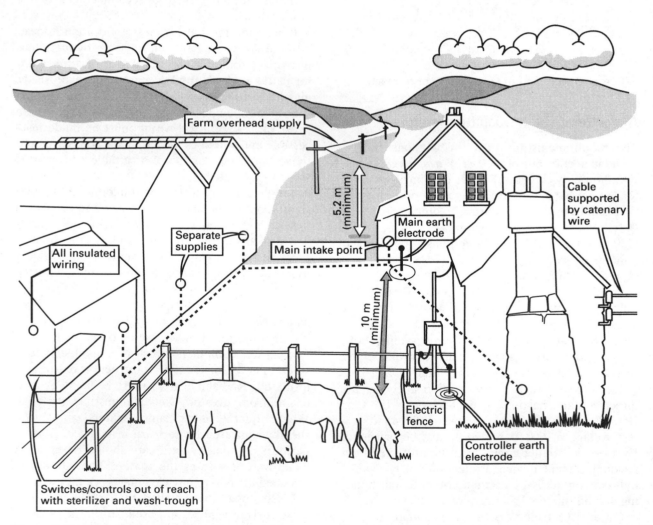

Figure 4.14  Some farm requirements

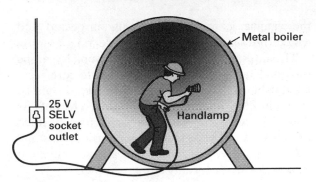

Figure 4.15 Using a hand-held lamp in a restrictive conductor location

installations within or installations intended to supply electrical equipment in confined spaces surrounded by conducting material (see Figure 4.15). They do not apply to any location where freedom of movement is possible. The protective measures against both direct and indirect contact can be provided by SELV or FELV provided that the voltage does not exceed 25 V a.c. or 60 V d.c. ripple-free. Protection against direct contact will need to be achieved with barriers or enclosures to IP2X or IPXXB or insulation capable of withstanding a test voltage of 500 V for 60 s. The protective measures, obstacles and placing out of reach are not allowed.

For the protection against indirect contact, SELV, FELV, general earthing and bonding, electrical separation and use of Class II equipment are all recognised. A supply to a handlamp or a socket outlet intended to supply a handlamp has to be protected by SELV. A similar requirement is for using an electrical hand-held tool. Where equipment requires a functional earth, then FELV can be used. Unless the means of isolation is part of the fixed installation within the restrictive conductor location, it must be situated outside for immediate access.

## Equipment having high earth leakage current

The requirements for such equipment are dealt with in Section 607 of the *IEE Wiring Regulations*. The scope of the Regulations refer to BS 7002 *'Specification for safety of information technology equipment including electrical business equipment'* (EN 60 950: 1988). Where the leakage current from such equipment does not exceed 3.5 mA, no special precautions are necessary.

Many Class I computers today incorporate filters

that produce leakage currents of around 2–3 mA and although not dangerous it is essential for the earthing of the equipment to be properly carried out. It is often the final connection from a socket outlet to the equipment that becomes faulty causing the metal casing of the equipment to become live. For this reason the earthing of the equipment should be regularly checked. Where a residual current device is used to protect more than one piece of stationary equipment and the earth leakage current is in excess of 3 mA it has to be verified that the total leakage current does not exceed 25% of the r.c.d. tripping current.

For items of equipment having leakage currents known to exceed 3.5 mA but not 10 mA, such equipment should either be permanently connected to the fixed wiring or connected via BS EN 60309–2 plugs and sockets. Where equipment exceeds 10 mA leakage in normal use then it has to be permanently connected to the fixed wiring.

The only provision for using a BS EN 60309–2 plug and socket is that the protective conductor in the flexible cable supplying the equipment is supplemented by a separate 4 mm$^2$ conductor and there must be an additional contact in the plug and socket outlet.

In fixed wiring installations, every final circuit intended to accommodate several items of stationary equipment, where it is known that the earth leakage current will exceed 10 mA, has to have a high integrity protective connection. There are several ways of achieving this: one is to provide a 10 mm$^2$ single protective conductor, another is to provide an earth monitoring device. An alternative approach is to use a 30 A ring final circuit and

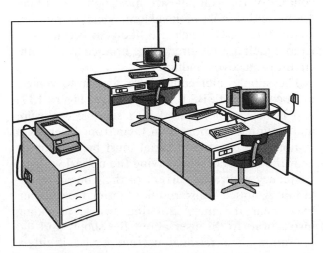

Figure 4.16 Data processing equipment

provided there are no spurs connected, the ends of the protective conductor (1.5 mm² minimum) are required to be connected into separate earth terminals at the fuseboard.

If the supply is from a TT earthing system and stationary items of equipment have an earth leakage exceeding 3.5 mA, the product of the earth leakage current and twice the value of the earth electrode resistance must not exceed 50. Where this cannot be achieved a double-wound transformer or equivalent device must be used.

## Caravan vehicle and caravan site installations

These electrical installations apply to trailer leisure accommodation vehicles and self-propelled leisure accommodation vehicles as well as site requirements. Section 608 of the *IEE Wiring Regulations* cover these requirements, only dealing with supplies up to 250 V/440 V.

Protection against direct contact does not include obstacles or placing out of reach and protection against indirect contact does not include non-conducting locations, earth-free equipotential bonding and electrical separation. Where the earthing arrangement of a vehicle is likely to be by automatic disconnection of the supply and a residual current device is to be provided, the wiring system must include protective conductors to the inlet coupler, the exposed conductive parts of electrical equipment and the protective contacts of socket outlets. If the protective conductor is not incorporated within the enclosure of the cable conduit or trunking, its size must be not less than 4 mm² and it must be insulated. Metal extraneous conductive parts which are likely to become live from a fault are required to be bonded by a 4 mm² circuit protective conductor.

The supply inlet coupler for a caravan vehicle must comply with BS EN 60309–2 (see Figure 4.17) and be accessible, placed in a suitable enclosure and mounted not more than 1.8 m from the ground. A notice of durable material must be fixed on or near the inlet recess specifying the normal voltage, frequency and rated current of the caravan installation. A notice is also required inside the caravan, fixed near the main isolating switch, providing instructions to the user about the supply and the operation of a residual current device if fitted. The notice should also provide information when the vehicle is to be periodically inspected and tested (not less than once every 3 years).

The internal wiring of a caravan vehicle can be sheathed flexible cables, flexible single-core insulated cables in PVC conduit or stranded insulated cables in PVC conduit. Cable sizes must be not less than 1.5 mm² and cable supports for sheathed cables must not exceed 0.4 m for vertical runs and 0.25 m for horizontal runs. No electrical equipment is to be located near any storage of fuel.

In terms of lighting, luminaires should be fixed directly to the structure (or lining) of the vehicle and the mounting should allow a free circulation of air to flow between the body of the luminaire and the vehicle's structure. Any luminaire that has a dual voltage operation must be fitted with separate lampholders for each voltage and must be designed so that lamps of different voltage cannot be wrongly inserted. The design must also consider any possible damage from heat if both lamps are on at the same time as well as provide effective separation of the internal wiring from the difference in voltage. Any accessories used that are subject to the effects of moisture must be designed to IP55.

Caravan pitch supply equipment should preferably be connected by underground cable and, unless otherwise protected from mechanical damage, be outside any caravan pitch or other area where tent pegs and ground anchors may be driven (see Figure 4.17). Where overhead conductors are used, they should be suitably constructed and located 2 m outside the vertical surface of any caravan pitch and at a height not less than 6 m in the vehicle movement area (in other areas 3.5 m). The supply equipment must be installed adjacent to the pitch and not more than 20 m from any point on the pitch which it is intended to serve.

A socket outlet which is part of the pitch supply equipment must be double pole and comply with BS EN 60309–2 (key position 6h) and be protected by IPX4. It should be placed at a height of between 0.8 m and 1.5 m and be of 16 A rating. It must be protected by an overcurrent device. Socket outlets can be protected individually or in groups of not more than three, by a residual current device complying with BS 4293. Grouped socket outlets must be on the same phase and socket outlets must not be bonded to any PME terminal. Where a TN–C–S earthing system is provided the protective conductor of each socket outlet has to be connected to an earth electrode.

A caravan site should be inspected and tested

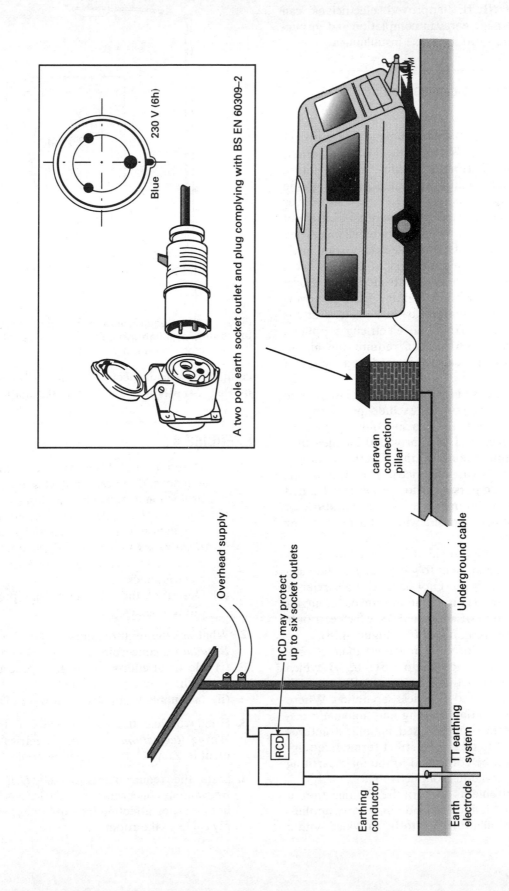

230 V (6h)

Blue

A two pole earth socket outlet and plug complying with BS EN 60309-2

Overhead supply

RCD may protect
up to six socket outlets

caravan
connection
pillar

Underground cable

Earthing
conductor

RCD

TT earthing
system

Earth
electrode

Figure 4.17  Supply to a caravan

every year. NICIEC-approved electricians can obtain a two-page caravan completion and inspection certificate form for these installations.

## Highway Power supplies and street furniture

These installations have to comply with Section 611 of the *IEE Wiring Regulations*. The Regulations cover highway distribution circuits, street furniture and other street located equipment (street lighting and illuminated signs, etc.). They may also cover similar equipment used by the public which is not designated as a highway or part of a building. They do not apply to a regional electricity company's work. It is worth mentioning that whereas most public lighting is usually controlled by a Local Lighting Authority some lighting such as advertising signs, telephone kiosks, bus shelters, etc. is controlled directly from the electricity supplier's cut-out. Regulation 611–03–02 requires the supplier's approval if the cut-out is used as a means of isolation.

The measures used for protection against electric shock from direct contact exclude protection by obstacles but allow the placing out of reach of low voltage overhead lines provided the measures meet the requirements of the *Electricity Supply Regulations 1988* (as amended). The Regulations also allow skilled persons who are specially trained to carry out maintenance of street furniture or equipment that is located within 1.5 m of a low voltage overhead line.

Where a door is provided in street furniture or street located equipment to gain access to the electrical equipment, is must not be used as a barrier or enclosure. There has to be an intermediate barrier to give protection of at least IP2X which can only be removed by using a tool (see Figure 4.18).

The measure of protection against indirect contact is covered in Regulation 611–02–03 which excludes non-conducting locations, earth-free equipotential bonding and electrical separation. Where earthed equipotential bonding and automatic disconnection of the supply is used, metallic structures (not connected to or part of street furniture/equipment) must not be connected to the main earthing terminal as extraneous conductive parts.

Other requirements concern the identification of cables, external influences and temporary supplies. Records of installations are to be provided with a

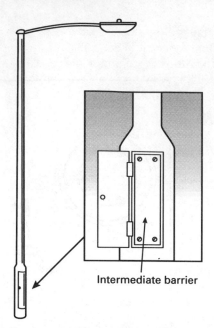

**Figure 4.18** Requirements for gaining access to the supply of a lighting column: removal of the internal cover requires a key or tool

completion and inspection certificate as required by Reg. 741–01–01 of the Wiring Regulations.

## EXERCISE 4.2

1. (a) Describe the difference between direct contact and indirect contact as measures of protection against electric shock.
   (b) List FIVE techniques recognised by Regulation 8 of the *Electricity at Work Regulations* that are used as suitable precautions to safeguard against the risk of electric shock.
   (c) Describe the measure of protection offered by IP44.

2. What are the requirements of the *IEE Wiring Regulations* concerning:
   (i) electrical equipment in the space under a bath;
   (ii) the supply to an electric towel rail?

3. Briefly outline the requirements of the *IEE Wiring Regulations* regarding electrical equipment in Zone C of a swimming pool.

4. State the requirements of the *IEE Wiring Regulations* concerning the restrictions inside a hot air sauna affecting the equipment and supply cable to the room

5. A horse riding stable is provided with a TT earthing system. What considerations should be given to the position of the earth electrode and what special requirements apply to socket outlets in the building?

6. (a) What is the maximum a.c. voltage which can be used for supplying handlamps in a restrictive conductive location?
   (b) State THREE locations where the *IEE Wiring Regulations* might apply to these locations.

7. With reference being made to the *IEE Guidance Notes, No. 5*, 'Protection against electric shock', what are the permissible earth leakage currents for:
   (i)   Class I portable appliances;
   (ii)  Class I heating appliances;
   (iii) Class II appliances.

8. Distinguish between the following terms
   (i)   caravan park and caravan pitch
   (ii)  Class I equipment and Class II equipment;
   (iii) SELV and FELV;
   (iv)  Street furniture and street located equipment.

9. Draw a neatly labelled diagram to show how a 240 V, single-phase distribution board is internally connected to supply three caravan BS EN 60309–2 socket outlets.

10. (a) Briefly explain the purpose and requirements associated with public lighting.
    (b) List a number of important points when erecting and installing lighting columns.

# Assessments

---

## Objectives

After working through this chapter you should be able to:

1  *apply your knowledge and understanding of:*
   - *communications and industrial studies*
   - *selection and installing wiring systems*
   - *special installations and locations*
   - *inspecting, testing and commissioning work;*

2  *apply your knowledge and understanding of topics relevant to electrical theory with the aid of reference material.*

## Multiple-choice questions

### *Communications and industrial studies*

**01.** Which one of the following is NOT a statutory instrument?
a) *Electricity at Work Regulations, 1989*
b) *IEE Wiring Regulations, 1992*
c) *Noise at Work Regulations, 1989*
d) *Electricity Supply Regulations, 1988.*

**02.** The two bodies known as the HSC and HSE were established under the:
a) *Health and Safety at Work etc. Act, 1974*
b) *Offices, Shops and Railway Premises Act, 1963*
c) *Factories Act, 1961*
d) *Public Health Act 1961.*

**03.** The British Standard BS 7671: 1992 is concerned with the:
a) requirements for electrical installations
b) safe use of ladders, step ladders and trestles
c) avoidance of danger from overhead electric lines
d) electrical test equipment for use by electricians.

**04.** The regulatory/statutory instrument that requires precautions to be taken against the risk of death or personal injury from electricity is called the:
a) *Building Regulations*
b) *IEE Wiring Regulations*
c) *Electricity at Work Regulations*
d) *Electricity Supply Regulations.*

**05.** The regulatory/statutory instrument concerned with declaration of supply phases, frequency and voltage is called the:
a) *Building Regulations*
b) *IEE Wiring Regulations*
c) *Electricity at Work Regulations*
d) *Electricity Supply Regulations.*

**06.** The regulatory/statutory instrument whose purpose is to protect persons, property and livestock against hazards arising from an electrical installations is called the:
a) *Building Regulations*
b) *IEE Wiring Regulations*
c) *Electricity at Work Regulations*
d) *Electricity Supply Regulations.*

**07.** Which one of the following buildings is NOT EXEMPT from Building Regulations?
a) public library
b) mobile home
c) greenhouse
d) barn.

**08.** The submission of a 'tender' to carry out work at a stated price is called an:
a) agreement
b) acceptance
c) offer
d) estimate.

**09.** The scale used for a site plan of a 'proposed' building is normally:
a) 1: 1000
b) 1: 500
c) 1: 100
d) 1: 50

**10.** An 'as fitted' drawing is one that provides information on the:
a) location of equipment using BS 3939 symbols
b) location of fixed electrical appliances
c) actual electrical work carried out
d) fixed central heating and plumbing work.

**11.** A 'bills of quantities' is a method of tendering that ensures all estimates:
a) are legally binding
b) are prepared with the same information
c) cover the cost for unforseen work
d) exclude unit costs.

**12.** What type of diagram is shown in Figure 5.1?
a) circuit diagram
b) wiring diagram
c) schematic diagram
d) block diagram.

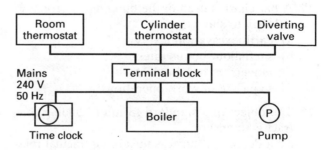

Figure 5.1   Central heating system

**13.** With reference to the *Electricity at Work Regulations*, Regulation 12 is concerned with:
 a) earthing
 b) competence
 c) equipment
 d) isolation.

**14.** A written authorization to carry out activities on or about a 'live' installation is called a:
 a) completion certificate
 b) permit to work
 c) variation order
 d) contract instruction.

**15.** If you have a grievance over some matter at work, your first action is to notify:
 a) your immediate supervisor
 b) the regional union representative
 c) an independent advisor
 d) the body called ACAS.

**16.** For an architect to see if building permission can be granted, he/she must make a request to his/her local authority for:
 a) outline planning
 b) Building Regulations
 c) full planning
 d) inspection certificates.

**17.** Which person listed below is likely to be called the architect's site inspector?
 a) quantity surveyor
 b) structural engineer
 c) clerk of works
 d) main contractor.

**18.** In building terms, the alteration or modification of any work shown on a contract drawing is called a:
 a) bill of quantity
 b) provisional sum
 c) variation
 d) prime cost sum.

**19.** A bar chart is used in the building construction industry to show the:
 a) earth excavation levels
 b) site labour bonus rates
 c) movement of labour on site
 d) sequence of work operations.

**20.** The object of a standard form of building contract is to ensure:
 a) a degree of fairness in sub-contractual relations

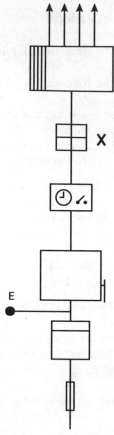

Figure 5.2   Heating circuit supply

 b) insurance protection is provided for site operatives
 c) statutory and regulatory requirements are met on site
 d) no disputes can arise between interested parties.

**21.** What is the name of the BS 3939 installation graphical location symbol marked X shown in Figure 5.2?
 a) Energy meter
 b) Contactor
 c) Controller
 d) Fan.

**22.** A weekly record of an operative's work activities on site is called a:
 a) job sheet
 b) day work sheet
 c) time sheet
 d) take-off sheet.

84

**23.** In the electrical installation industry, the subject called: *Installing electrical systems and equipment*, leads to the award of NVQ Level:
a) I
b) II
c) III
d) IV.

**24.** Which of the following is the lead body for the electrical installation engineering industry?
a) ETA
b) NICEIC
d) EIEITO
e) JIB.

**25.** Under the Employment Protection Act, for sixteen or more hours of working, an employer has to give an employee a contract of employment. This has to be:
a) immediately upon starting work
b) at the discretion of the employer
c) within 13 weeks of starting work
d) after the employee's first wages.

## *Selecting and installing wiring systems*

**26.** The cable term 'LSF' means:
a) light sheath feature
b) low smoke fume
c) limited size fitting
d) long strand fibre.

**27.** The internal conductors of a rising main busbar trunking system used in a multi-storey building must be:
a) securely fixed to allow no movement
b) arranged to expand and contract
c) spaced apart a minimum of 50 mm
d) supported every 1 m of length.

**28.** PVC cables should not be handled in temperatures:
a) between 20°C and 30°C
b) between 10°C and 20°C
c) between 0°C and 10°C
d) less than 0°C

**29.** A suitable wiring system chosen for its aesthetic appearance inside a church would be:
a) MICC cable
b) uPVC trunking
c) PVC/SWA cable
d) metal conduit.

**30.** A 'tapping box' is an accessory associated with:
a) PVC conduit
b) metal conduit
c) MIMS cables
d) busbar trunking.

**31.** The purpose of segregating cables of different category in trunking is to reduce the risk of:
a) fire and spread of smoke
b) electromagnetic interference
c) mechanical damage
d) wrongful connection.

**32.** In the *IEE Wiring Regulations*, the letters 'AD' are used to denote the following external influence:
a) presence of water
b) presence of fauna
c) building design
d) mechanical stress.

**33.** Non-armoured PVC cables for general use are normally insulated to a voltage grade of:
a) 250/415 V
b) 300/500 V
c) 600/1000 V
d) 1000 V.

**Note:** Some of the following questions require students to refer to the *IEE Wiring Regulations* or *IEE On-Site Guide*.

**34.** The maximum continuous operating temperature for XLPE armoured cables is:
a) 100°C
b) 90°C
c) 60°C
d) 30°C.

**35.** The factor to be applied to MI cable to determine the minimum internal radius of the bend is:
a) 10
b) 8
c) 6
d) 3.

**36.** One advantage of installing an overhead busbar trunking in a factory is that there is no need to be concerned with the:
a) running of separate protective conductors
b) requirements for mechanical protection
c) final layout of connected machinery
d) problems of voltage drop.

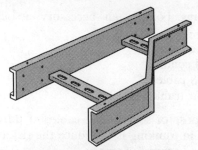

Figure 5.3   Cable ladder accessory

37. The cable ladder accessory shown in Figure 5.3 is called a:
   a) concentric reducer
   b) left-hand reducer
   c) two-way intersection
   d) through intersection.

38. The maximum length of span for a metal conduit between buildings is:
   a) 6 m
   b) 5 m
   c) 4 m
   d) 3 m.

39. The maximum horizontal spacing of supports for PVC trunking, exceeding 300 mm and not exceeding 700 mm is:
   a) 2.0 m
   b) 1.5 m
   c) 0.5 m
   d) 0.3 m.

40. What size metal conduit is required to contain six 2.5mm$^2$ and eight 1.5 mm$^2$ single-core PVC cables when its length of run is 4.5 m and it has one bend?
   a) 32 mm
   b) 25 mm
   c) 20 mm
   d) 16 mm.

41. What size plastic trunking is required from Table 5F of the *IEE On-site Guide* to contain the following stranded single-core copper cables: twenty 1.5 mm$^2$, twenty-four 2.5 mm$^2$, twelve 6 mm$^2$ and ten 10 mm$^2$?
   a) 75 mm × 37.5 mm
   b) 75 mm × 25 mm
   c) 50 mm × 50 mm
   d) 50 mm × 37.5 mm.

42. Which one of the following is a property of a good conductor?

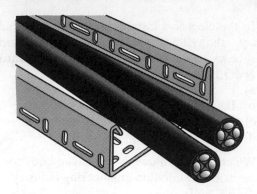

Figure 5.4   Armoured cables on perforated cable tray
   a) high resistivity
   b) low current density
   c) high conductivity
   d) low tensile strength.

43. Which one of the following is the correction factor for a BS 3036 semi-enclosed fuse?
   a) 0.725
   b) 0.537
   c) 0.375
   d) 0.257.

44. The tables 4D1 to 4L4 of the *IEE Wiring Regulations* are based on an ambient temperature of:
   a) 35°C
   b) 30°C
   c) 25°C
   d) 20°C.

45. With reference to Table 4A of the *IEE Wiring Regulations*, the installation method shown in Figure 5.4 is:
   a) Method 11
   b) Method 12
   c) Method 18
   d) Method 19.

46. With reference to Tables 54B to 54F of the *IEE Wiring Regulations*, the value of '$k$' for an aluminium protective conductor used as armouring in an XLPE cable is:
   a) 125
   b) 116
   c) 94
   d) 85.

47. The phase conductor of a circuit is 50 mm$^2$. The minimum size of the corresponding protective conductor that can be used from Table 54G of the *IEE Wiring Regulations* is:

a) 50 mm²
b) 45 mm²
c) 30 mm²
d) 25 mm².

**48.** With reference to Table 54G (Wiring Regs), what is the most suitable size light-duty metal conduit that can be used as a protective conductor if it contains a 50 mm² copper phase conductor?
a) 32 mm dia, (116 mm²)
b) 25 mm dia, (89 mm²)
c) 20 mm dia, (59 mm²)
d) 16 mm dia, (47 mm²).

**49.** With reference to Regulation 543–01–03 of the Wiring Regulations and given that $I = 400$ A, $t = 0.035$ s and $k = 115$, the value of $S$ is:
a) 1.00 mm²
b) 0.82 mm²
c) 0.75 mm²
d) 0.65 mm².

**50.** With reference to Table 4D2B of the Wiring Regulations, a 4 mm² two-core cable has 11 mV/A/m. If the cable carries a current of 20 A, what is the maximum length that can be used to keep within a voltage drop limit of 9.6 V?
a) 50.55 m
b) 43.64 m
c) 23.98 m
d) 16.97 m.

## *Inspecting, testing and commissioning work*

**51.** The label on a BS 951 earth clamp should state:
a) Danger electricity – do not remove
b) Safety electrical connection – do not remove
c) Consumer earth – do not remove connection
d) Electrical earth connection – do not remove.

**52.** The equipment used in a dusty environment should be designated with the code:
a) CA1
b) AD2
c) BE3
d) AE4.

**53.** Holes in a joist for cables to pass through should be drilled from the top or bottom of the joist at a distance of:

a) 100 mm
b) 75 mm
c) 50 mm
d) 35 mm.

**54.** All the following are assessment characteristics for a supply into a building obtained by enquiry from a regional electricity company, EXCEPT:
a) size of consumer's main switchgear
b) nature of supply current and frequency
c) type and rating of main supply cut-out fuse
d) prospective short circuit current at intake point.

**55.** Which of the following instruments is the most suitable for carrying out the verification of leakage current to earth?
a) Earth fault loop impedance tester
b) High reading ohmmeter
c) Earth electrode tester
d) Low reading ohmmeter.

**56.** A ring final circuit test of continuity is to verify that there are no conductor breaks in the ring and also no:
a) multiple loops
b) different size cables used
c) incorrect polarity
d) short circuit faults.

**57.** The minimum insulation resistance for a complete installation at a test voltage of 500 V d.c. is:
a) 5.0 MΩ
b) 1.0 MΩ
c) 0.5 MΩ
d) 0.2 MΩ.

**58.** When carrying out a visual inspection of a rising main trunking busbar system, which one of the following must be verified?
a) Final circuit voltage drop
b) Prospective short circuit current
c) Presence of fire barriers
d) Continuity of trunking through floors.

**59.** When a wiring system is exposed to the weather, all metal parts of the systems should be of a:
a) moisture repellant material
b) corrosion resistant material
c) copperclad, aluminium material
d) non-ferrous, magnetic material.

60. The term used to describe the effects on an installation in regards to its design and safe operation is called:
   a) maintainability
   b) compatibility
   c) external influence
   d) nature of supply.

61. Which one of the following is classified as a Category 3 circuit?
   a) telephone circuit
   b) data transmission circuit
   c) intruder alarm circuit
   d) emergency lighting circuit.

62. When inspecting and testing installations, all the following are under the control of a local authority conditions of licence, EXCEPT:
   a) domestic dwelling
   b) cinema
   c) leisure complex
   d) launderette.

63. In discovering a continuity test reading above 50 Ω between a stainless steel sink and an adjacent gas hob, the correct action to take would be to:
   a) carry out equipotential bonding
   b) retest the circuit with an insulation tester
   c) earth each extraneous part separately
   d) leave the two extraneous parts isolated.

64. A 30 mA r.c.d when subject to a test of 150 mA, should trip out in:
   a) 200 ms
   b) 100 ms
   c) 50 ms
   d) ≤ 40 ms.

65. Domestic dwellings should be re-tested after an interval of:
   a) 15 years
   b) 10 years
   c) 5 years
   d) 3 years.

66. When making an insulation resistance test to earth on a completed installation, it is best practice to:
   a) place all switches in the off position
   b) remove all fuse links in place
   c) short the connection between neutral and earth
   d) leave all lamps in their lampholders.

67. The maximum earth fault loop impedance for a cooker circuit having no 13 A socket outlet in the control box but protected by a 45 A BS 1361 fuse is:
   a) 1.0 Ω
   b) 0.6 Ω
   c) 0.3 Ω
   d) 0.1 Ω.

68. On inspecting the suitability of BS 1361 Type I fuses used in a fuseboard, these devices have a rated breaking capacity of:
   a) 33.0 kA
   b) 18.0 kA
   c) 16.5 kA
   d ) 0.5 kA.

69. Which of the following is the British Standard number for emergency lighting?
   a) BS 7671
   b) BS 5839
   c) BS 5499
   d) BS 5266.

70. In a high risk task area, such as a control area, CEN 169 WG3 requires emergency lighting to be activated within a time of:
   a) 5.00 s
   b) 2.00 s
   c) 0.50 s
   d) 0.25 s.

71. When inspecting and testing an emergency lighting system, a one-hour test is required every:
   a) 12 months
   b) 6 months
   c) 3 months
   d) 1 month.

72. Manual call points in a fire alarm system should be sited in a well-lit unobstructive position at a mounting height of:
   a) 1.75 m
   b) 1.55 m
   c) 1.40 m
   d) 1.35 m.

73. Where sleeping people are to be woken by a fire alarm sounder, the minimum sound level at the bed-head should be:
   a) 75 dBA
   b) 65 dBA
   c) 50 dBA
   d) 40 dBA.

**74.** A TN–C–S earthing system is forbidden to be used in a:
a) petrol filling station
b) private dwelling
c) agricultural premises
d) caravan park.

**75.** A hazardous area in a petrol filling station in which an explosive gas/air mixture is not likely to occur in normal operation is called:
a) Zone 3 area
b) Zone 2 area
c) Zone 1 area
d) Zone 0 area.

## Special installations or locations

**76.** Electrical equipment may only be installed under a bath if:
a) it is connected to the main earth terminal
b) access to it is by the use of a tool
c) a SELV supply is used
d) a FELV supply is used.

**77.** The minimum distance that one can install a socket outlet in a shower room is:
a) 3.0 m
b) 2.5 m
c) 2.0 m
d) 1.5 m.

**78.** Where a corded telephone point is installed in a bathroom the recommended distance from the bath is:
a) 2.5 m
b) 2.0 m
c) 1.5 m
d) 0.5 m.

**79.** Luminaires installed within 2.5 m of a bath must be non-metallic and be either totally enclosed or have:
a) a Home Office type skirt fitted.
b) a mass not exceeding 3 kg
c) an IP4X index of protection
d) a 5 mm drainage hole.

**80.** The minimum height at which electrical equipment should be installed above a diving board over a swimming pool is:
a) 3.5 m
b) 3.0 m
c) 2.5 m
d) 2.0 m.

**81.** Electrical equipment installed in Zone A of a swimming pool should be designed with a protection afforded by:
a) IPX8
b) IP42
c) IPX2
d) IP11.

**82.** The external influence classified as *AF* refers to:
a) vibration
b) induction
c) immersion
d) corrosion.

**83.** All electrical equipment used in a hot air sauna must be designated:
a) IPX2
b) IPX8
c) IP12
d) IP24.

**84.** On a construction site, the preferred a.c. voltage for a portable hand lamp used in a damp location should be:
a) 110 V
b) 64 V
c) 50 V
d) 25 V.

**85.** On a construction site having TN–S earthing, the maximum disconnection time for protective devices supplying 110 V hand held tools is:
a) 5.00 s
b) 0.40 s
c) 0.35 s
d) 0.25 s.

**86.** On a construction site having TN–S earthing, the maximum earth fault loop impedance for a final circuit protected by a Type B, BS 3871 16 A m.c.b is:
a) 8 $\Omega$
b) 6 $\Omega$
c) 5 $\Omega$
d) 3 $\Omega$.

**87.** Where an r.c.d. is used on a construction site, fed from a TN earthing system, the product of the earth fault loop impedace and rated residual operating current of the r.c.d. must be equal to or less than:
a) 100 V
b) 50 V
c) 25 V
d) 12 V.

**88.** All plugs and socket outlets used on a contruction site should comply to:
a) BS 4343
b) BS 3939
c) BS 2632
d) BS 1986.

**89.** In the absence of manufacturers' guidance, the minimum distance a radiant heater can be installed near livestock on a farm is:
a) 2.0 m
b) 1.5 m
c) 1.0 m
d) 0.5 m.

**90.** The maximum number of socket outlets that can be protected by an RCD in a caravan pitch is:
a) 10
b) 6
c) 3
d) 2.

**91.** A caravan site should be periodically inspected and tested every:
a) 12 months
b) 6 months
c) 3 months
d) 1 month.

**92.** A petrol filling station should be periodically inspected and tested every:
a) 12 months
b) 6 months
c) 3 months
d) 1 month.

**93.** The maximum period of inspection and testing of the following installations is 3 years, EXCEPT for a:
a) theatre
b) emergency lighting
c) industrial premises
d) caravan.

**94.** On a farm an electric fence can be controlled by:
a) 4 controllers
b) 3 controllers
c) 2 controllers
d) 1 controller.

**95.** When r.c.d. is used in a TT earthing system on a farm, the product of the earth fault loop impedance and the r.c.d.'s tripping current must NOT exceed:

a) 50 V
b) 25 V
c) 12 V
d) 5 V.

**96.** In a restrictive conductive location, a supply to a socket outlet feeding hand-held tools is required to be protected by:
a) SELV or earthing and bonding
b) electrical separation or SELV
c) SELV or obstacles
d) placing out of reach or SELV.

**97.** In a location intended to serve several items of computer equipment with a possible leakage to earth exceeding 10 mA, the circuits shall be provided with a:
a) BS 88 Part 2 cartridge fuse
b) high integrity earth conductor
c) 30 mA residual current device
d) centre-tapped transformer.

**98.** A SELV system is a system that requires:
a) no earthing facility
b) a d.c. supply of 120 V
c) an a.c. supply of 110 V
d) restrictive access.

**99.** In highway power supply, all the following protective measures against indirect contact are not permitted EXCEPT:
a) electrical separation
b) earth free equipotential bonding
c) earthing and bonding
d) non-conducting location.

**100.** The maximum disconnection time for protective devices in circuits supplying fixed highway power supplies is:
a) 5.0 s
b) 0.4 s
c) 0.2 s
d) 0.1 s.

---

## Written questions

**Q1.** (a) Explain why a temporary electrical installation on a building site should be designed to at least the same standard as a permanent installation.

(b) For a temporary installation, state several requirements of the *IEE Wiring Regulations* for (i) types of switches, plugs and sockets to be used, (ii) use of overhead cables, and (iii) frequency of testing.

(c) State ONE example in each case of plant or equipment connected to supplies of 25 V, 110 V, 230 V and 400 V.

**Q2.** (a) Describe some of the regulatory requirements concerned with earthing and bonding clamps.

(b) Describe some of the requirements of the *IEE Wiring Regulations* concerned with residual current devices.

**Q3.** (a) What is meant by the term 'earth fault loop impedance'?

(b) Why is it essential for a final circuit to have a low earth fault loop impedance?

(c) What is the maximum disconnection time for a protective device connected in a 400 V a.c. circuit feeding portable equipment intended for manual movement?

(d) Explain the difference between the terms 'overload current' and 'short-circuit current'.

(e) What is the maximum value of external earth loop impedance allowed for a TT earthing system?

**Q4.** (a) A 2-core PVC armoured cable supplies 230 V equipment, having exposed conductive parts, outside the main equipotential zone of a building. If the circuit protective device is a 50 A BS 88 Part 2 fuse which affords indirect contact protection, state the maximum permitted disconnection time.

(b) From Appendix 3 of the Wiring Regulations, determine the minimum earth fault current which satisfies the required disconnection time.

(c) Find from the appropriate table in the *IEE Wiring Regulations*, the maximum value of earth fault loop impedance and use this value to determine the external earth fault loop impedance, given that $R_1$ and $R_2 = 0.28 \ \Omega$

**Q5.** (a) Describe the internal components of a metal-clad six-way consumer unit fitted with overcurrent protection and earth leakage protection.

(b) State the types and uses of various types of miniature circuit breaker.

(c) Make a sketch to show how different types of m.c.b. compare with each other and determine their tripping currents for a device rated at 10 A.

**Q6.** (a) Explain the dangers associated with a 'borrowed' neutral conductor.

(b) State the Wiring Regulations' requirements for the wiring of final circuits in regards to 'borrowed' neutral conductors.

**Q7.** Figure 5.5 shows a motor control circuit fed from a distribution board using different wiring systems.

(a) Draw a neatly labelled diagram of the circuit.

(b) Describe the procedure for isolating the supply to reverse the motor's direction of rotation.

**Q8.** List several important practical design considerations for:

(a) an emergency lighting system.

(b) a fire alarm system.

**Q9.** List some of the important consideration required in a maintenance manual.

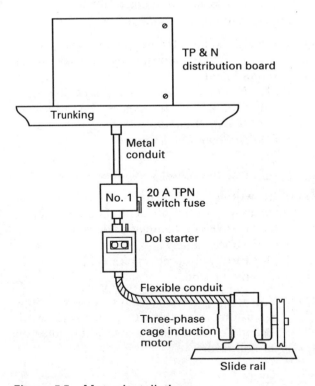

Figure 5.5    Motor installation

**Q10.** You have been called out to a factory to see why a motor is not running. If the motor is a three-phase induction motor fed from a distribution board, what procedure would you follow to find the fault?

**Q11.** You have just obtained a contract to do electrical work in a garage premises. Describe the method of providing information to the client, stressing the importance of safety in terms of using handlamps, lighting in an inspection pit and spray booth, and a socket outlet supplying a high pressure cleaner outdoors.

**Q12.** On a construction site, you are asked by the main building contractor if it safe to construct the site office underneath high voltage power cables. Describe the steps that should be taken to avoid danger.

**Q13.** Briefly explain the European requirements for emergency lighting of premises in new buildings.

**Q14.** Briefly explain the following terms:
(1) legislation
(2) civil law
(3) criminal law
(4) European community law
(5) statutory instrument.

**Q15.** Briefly explain the scope and purpose of TWO of the following regulatory/statutory requirements:
(1) *Health and Safety at Work etc Act*
(2) *Control of Substances Hazardous to Health Regulations*
(3) *Building Regulations*
(4) *IEE Wiring Regulations*
(5) *Fire Regulations*.

**Q16.** In writing an electrical specification for cables, draft brief notes on the following:
(1) PVC/PVC sheathed cable routes
(2) MIMS cables
(3) PVC armoured cables laid directly in the ground.

**Q17.** You are requested by your company to show a customer around the electrical work recently carried out on site. Write notes on the following:

(1) site safety and site rules
(2) technical knowledge of work
(3) attitude and behaviour

**Q18.** You and several other electrical apprentices are instructed to off-load from a lorry, a very large switchboard. The switchboard then has to be moved on firm flat ground to its destination. Explain how you would achieve this task if no lifting gear is available. In your answer, state numerous safety instructions that you and your helpers would need to consider.

**Q19.** Briefly describe the procedure for the following:
(a) treatment of:
  (i) electric shock
  (ii) burns from hot pitch
  (iii) bleeding wound;
(b) evacuation of an occupied premises.

**Q20.** Figure 5.6 shows the layout of a factory's main switchboard supplied at 400 V, three-phase, 4-wire.
(a) Draw a fully labelled diagram of the circuit.
(b) Show with a line diagram how fuse discrimination is achieved using BS 88 Part 2 fuses.

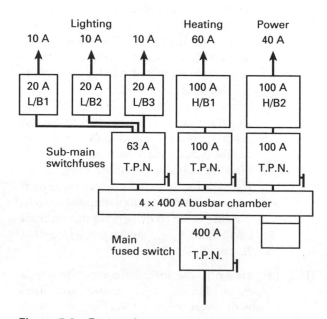

Figure 5.6   Factory layout

# Appendix

**Specimen form**

---

*Town and Country Planning Act 1971*

*Application form for permission to develop land*

Date received

Application No . ................................................................

1. Applicant (in block capitals)

Agent (to whom correspondence should be sent)

Name ........................................................................
Address ....................................................................
..................................................................................
..................................................................................
Tel. No. ....................................................................

Name ........................................................................
Address ....................................................................
..................................................................................
..................................................................................
Tel. No. ....................................................................

2. (a) Full address of location/land to which this application relates
Address .....................................................................................................................

   (b) Brief description of proposed development including the purpose(s) for which the land and/or buildings are to be used
.........................................................................................................................................................
.........................................................................................................................................................
.........................................................................................................................................................

If residential development, state number of dwelling units proposed and type if known (e.g. flats, houses bungalows etc.)
.........................................................................................................................................................
.........................................................................................................................................................

   (c) Area of site on which this application is made (hectares/acres)
.........................................................................................................................................................

   (d) State whether or not applicant owns or controls any adjoining land and if so, give its location
.........................................................................................................................................................
.........................................................................................................................................................

   (e) State whether the proposal involves:

|  |  |  |
|---|---|---|
| (i) | New building (s) | Yes/No |
| (ii) | Alteration/extension | Yes/No |
| (iii) | Change of use | Yes/No |
| (iv) | Construction of a new access to a highway | Yes/No |
| (v) | Vehicular/pedestrian | Yes/No |
| (vi) | Alteration of an existing access to a highway | Yes/No |

Note: Delete appropriately

3. Particulars of Application
State whether this application is for:

|  |  |  |
|---|---|---|
| (i) | Full planning permission | Yes/No |
| (ii) | Outline planning permission | Yes/No |

If 'Yes' to (ii) above, delete any of the following that are not reserved for subsequent approval:

      \* siting
      \* design
      \* external appearance
      \* means of access
      \* landscaping

   (iii) Approval of reserved matters following the grant of outline permission.    Yes/No

If 'Yes' to (iii) state the date and number of outline permissions
Date ....................................................................................
No. ......................................................................................

   (iv) Renewal of a temporary permission    Yes/No

*continued*

      (v)      Variation or removal of a condition
               on an existing permission                                         Yes/No

If 'Yes' to (iv) or (v), state the date and number of
previous permissions and identify the particular condition.
               Date ...............................................................................................
               No. ................................................................................................
               Condition .......................................................................................

4.  Particulars of present and previous use of buildings or land. State:
      (i)       Present use of buildings/land:
              .......................................................................................................
      (ii)     If vacant, the last previous use:
              .......................................................................................................

5.  Additional information
     (a)    Is the application for industrial, office, warehousing,
           storage or shopping purposes?                            Yes/No
           If the answer is 'Yes', complete Form Part 2 from
           the Planning Department.
     (b)    Does the proposed development affect any trees
           on the site?                                                  Yes/No
           If 'Yes', please indicate positions on the plan.
     (c)    (i)      How will surface water be disposed of?
                 .......................................................................................
          (ii)     How will foul sewage be dealt with?
                 .......................................................................................
     (d)    If new buildings are to be erected, or alterations/extensions to existing buildings are proposed, please state the
           materials and colours proposed for external finish of:
          (i)      Existing walls and roof...................................................................
          (ii)     Proposed walls and roof ................................................................

# Appendix

## Client/architect relations

After the consideration for a need to build, a client will appoint an **architect** to act for him to give opinion and advice and to discuss the terms of appointment for a design team of consultants. The architect will obtain from the client the necessary background information and receive instructions about the way the building project is to proceed. He will obtain an initial statement of the client's requirements such as the time-scale for completing the project and its financial involvement. He will also discuss with the client, preliminary details of the project, obtaining site plans, ordinance survey maps, etc. and make initial site visits.

The architect will call a meeting with the client and a **design team** consisting of quantity surveyor, civil, structural, mechanical and electrical engineers to establish responsibilities, prepare a plan of work and produce a timetable for the project. The design team will carry out studies on site and assist in the preparation of a feasibility report. With recommendations, it will be sent to the client for discussion and consideration.

At this **brief stage** the client may want to abandon, modify or issue new instructions. If he accepts the report, the architect and design team will then prepare a **directive** for outline proposals and scheme design.

This is the **sketch-plan** stage of the project and is to decide the general approach to layout, design and construction. It is done to obtain authoritative approval from the client on the outline proposals and accompanying report. The architect is required to develop the brief and carry out studies relating to the new work and similar completed work, trying out detail planning solution, etc., and deciding with the design team on general approach.

The consulting engineers and main contractor (if appointed) will provide the quantity surveyor with information for him to prepare an **outline cost plan**. This will be obtained from comparison of requirements of previous projects or from approximate quantities based on assumed specification. The architect will then compile dossiers on final sketch designs, recording all assumptions made and present a **written report** to the client.

To complete this stage, the architect has to clear up any outstanding matters with the client and then pass the information on to his design team. He will then prepare a full scheme design and report for the client and await his decisions and further instructions.

The next main stage in the project is to produce **working drawings** and the design team will gather information about all matters relating to the design specification, construction and cost. The team will prepare drawings, schedules, specifications, bills of quantities and tender documents. A list of main contractors (if not already selected) will be drawn up by the architect and he will complete all **documents** for obtaining tender.

The final stage in the project comes under the heading of **site operations** and is concerning:

(i)   project planning;
(ii)  operation on site;
(iii) project completion;
(iv)  feedback.

In project planning the architect will inform the client of the **contract conditions**, financial arrangements and methods of communicating instructions, etc. He will then send copies of the contract conditions to the main contractor for checking and signing. This will enable the contractor to programme the work according to the contract conditions and he can then appoint site staff and make reservation of early plant requirements.

The client has to sign the contract documents, arrange and prepare to hand over the site to the contractor. The architect will hold a **project meeting** with all parties. This will concern the date of handover of the site and the issuing of drawings and other production information. It will also include the nomination of subcontractors and suppliers, financial and insurance arrangements, and agree priorities, timetables, methods of programming and attendance at site meetings, etc.

Operation on site requires the architect to administer the terms of the contract. This will involve issuing AIs (**architect's instructions**) and certificates and nomination of any outstanding sub-contractors and suppliers. It will also involve claims and adjudication matters and the authorisation of 'daywork'. The client will want to be kept informed of progress, receive running financial statements and approve any increased costs.

**Site meetings** will be concerned with progress, the recording of work done and actions needed such as the position concerning site labour and supply of materials and site problems.

The completion of the building project is its handover to the client for occupation. The architect will inspect the building before the end the defect liability period, checking to see if all incomplete work is finally completed. He will then take action to remedy any defects, settle the final account and make sure that all the work is completed following the contract.

At the end, the project should be analysed for its management, construction and performance with respect to other similar projects.

**Note:** The above is the author's summary of the book *Plan of Work for Design Team Operation*, produced by RIBA Publications Limited. This can be purchased from the following address:

RIBA Publications,
Finsbury Mission,
39 Moreland Street,
London EC1V 8BB.

# Appendix

## Specimen form

---

### *Statement of terms and conditions of employment*

**Name of employer/employing body**

........................................................................................................................................
........................................................................................................................................
........................................................................................................................................

**General**

........................................................................................................................................
........................................................................................................................................
........................................................................................................................................

**Introduction**
This statement is issued in accordance with the terms of Part I of the Employment Protection (Consolidation) Act 1978, as amended, gives information about some of the principal terms and conditions of employment in ...............................
........................................................................................................................................

**Job Title and Duties**
Your grade will be that of 'Electrician' under the supervision of .............. You will be required to carry out work efficiently and economically and in accordance with the current IEE Wiring Regulations. You will be required to undertake flexible methods of working which are a condition of your employment and you must be fully acquainted with the Company's safety policy.

**Continuity of Employment**
Your previous employment with ................... will count as part of your continuous employment with this Company as commenced on .....................

**Probationary Period**
Not applicable

**Pay and Pay Statement**
On transfer from your previous employer to this Company, your salary will continue to be that which is appropriate to your grade. Your gross pay at the date of this statement is .............. per month which will be made up as follows:
    (i) Basic pay ................. and
    (ii) Flexible supplement ...........................

The flexible supplement is payable under an agreement with the recognised Trade Union for undertaking certain flexible methods of working and other conditions. It is not payable during sickness and certain other absences and only remains payable while the agreement is in force (see Company's industrial staff code obtained from ....................).
    You will be paid monthly, in arrears by credit transfer and with each payment you will receive a written statement showing gross pay, any allowances payable, deductions and net pay.

**Hours**
Your normal working hours, which are the conditioned hours for your grade, will continue to be 37.5 net per week, Monday to Friday, (exclusive of unpaid meal breaks), as laid down in the Company's industrial staff code. Daily attendance will not be more than eight hours during any consecutive nine hours between 7.30 a.m. and 6.30 p.m.

**Overtime**
You may be required to work extra hours beyond your normal hours at the request of the Company when the workload makes this necessary. Payment of overtime and overtime rates are laid down in the Company's industrial staff code. Premiums shall be paid at time-and-a half. For all hours worked between 1 p.m. on Saturday and normal starting time on Monday premiums shall be paid at double time.

**Annual Leave**
In addition to public and privilege holidays which total 8 days per year, your annual leave allowance will continue to be 21 days with pay. The leave year will run from 1 April to 31 March.

*continued*

## Sick Leave

You are required to notify and provide evidence of incapacity to work to the Company when you are absent because of sickness. The rules and regulations relating to notification, evidence of incapacity, any entitlement to sick leave and sick pay and other related matters are to be found in the Company's industrial staff code.

## Travel Allowance

You may be required to undertake official journeys, and this can include periods of temporary detached duty away from the Company. When travelling on authorised journeys, you will be reimbursed expenses and paid subsistence allowance in accordance with the Company's rules that are agreed with the recognised Trade Union.

## Superannuation

On joining the Company, you will be a member of the Company's Superannuation Scheme. Although membership is not compulsory, if you withdraw from the Scheme, you will be granted one further option to rejoin at a later date subject to age limits and evidence of good health or as may be determined from time to time in the Company's industrial staff code.

## National Insurance Contributions

A contracting-out certificate in respect of State Pension Scheme, is in force for this employment. You will be required to pay only the abated rates of the NI contributions on the upper tier of your earnings.

## Retirement

The retirement age for all staff is 60 years. For further details see the Company's industrial staff code.

## Notice

You are entitled to not less than the minimum period of notice stated below, unless you are summarily dismissed on disciplinary grounds and providing there has been at least four weeks continuous service:

    (i)      less than 5 years continuous employment – 5 weeks
    (ii)     For 5 years or more in continuous employment, the minimum period of notice is one week for each complete year, plus one week to a maximum of 13 weeks.

On resignation, employees are required to give not less than one week's notice. Further details can be found in the Company's industrial staff code.

## Safety Rules

You are required to observe the Company's safety rules and safety policy for the time being in force and any amendments thereto.

## Grievance Procedure

If you have a grievance relating to your employment, you should raise it orally with your immediate superior in the first instance. Thereafter, you should follow the procedure set out in the Company's industrial staff code.

## Discipline and Appeals Procedure

You will be responsible to the Director of the Company through his designated managers, in all matters affecting your work and you will be subject to, and required to comply with, such conditions of employment, rules of conduct and disciplinary procedures as may be announced from time to time. For further information see the Company's industrial staff code.

If you are dissatisfied with any disciplinary decision, you may appeal against the decision, in person or in writing to the appropriate person, following the arrangements set out in the Company's industrial staff code.

## Trade Union Membership

If you decide to join a recognised Trade Union, you are encouraged to play an active part and ensure your views are fully represented.

## Acceptance

If you accept the offer of employment you should sign one copy of these terms and conditions and return it to the undersigned at the address heading the accompanying letter. The other copy should be retained by you.

Signature on behalf of the employer ........................ Date ........................................................................

I accept the terms and conditions of employment with: ..................set out or referred to above.

Signature of employee........................................................ Date..................................

## Answers to exercises

### EXERCISE 1.2

1. (a) The measurements should be 10.2 m × 6.85 m
   (b) List the equipment as follows:

| Quantity | Description |
|---|---|
| 23 | Thorn Clipper 2, FCLV158/FCLD5 luminaires |
| 1 | Thorn 'Nova' NSB100 Gallery luminaire |

   (c) See Figure 1.9 and take plan and drop measurements. Assume the trunking is erected around the room.

2. (i) See Figure A4.1(a)
   (ii) See Figure A4.1(b)

3. (a) See Chapter notes
   (b) See Figure A4.2

4. There are really four main ingredients in the process of forming a contract between two parties, namely:

   (i)   the invitation to tender;
   (ii)  an offer;
   (iii) an acceptance;
   (iv)  a consideration.

   Briefly, in (i) the client or his agent sets out the proposed project details and invites contractors to tender for the work. In (ii) a contractor's quotation is produced for the work to be carried out. In (iii), this is the client's formal acceptance of the offer, and (iv) is the reward (usually the quoted price) for completing the work. Standard forms of contract ensure a degree of fairness in the contractual relationship between a client and the person (usually a contractor) carrying out the client's work.

Both parties are bound by various clauses, often established over many years and it is possible that a contract devised by a client will be written in his favour.

5. Consult manufacturers' catalogues and prices. Estimate the material and labour cost based on the assumption that you need a 12-way, three-phase fuseboard and the alteration takes 1 hour to complete.

6. In (i), it is important for the electrical contractor to provide a safe working environment for his operatives. A good site hut will provide the necessary accommodation for operatives to eat meals in, change clothing, wash and also provide shelter from adverse weather conditions. The hut should be kept dry and tidy and not used for storing bulk supplies of material and equipment or hazardous substances.

   In (ii), the HSW Act requires the provision of safe means of access and egress from the place of work. Suitable scaffolding, towers, ladders, trestles, etc., must be provided for work that cannot be done safely from the ground or part of the building structure. The electrical foreman

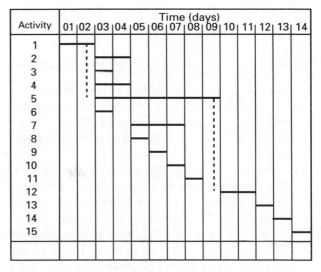

Figure A4.2   Bar chart for floodlighting project

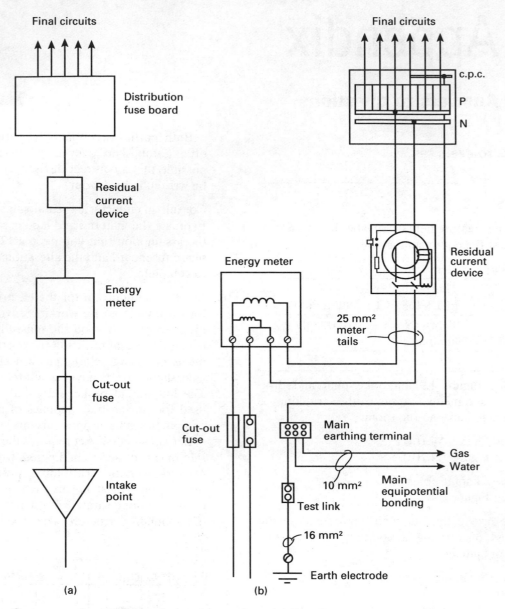

**Figure A4.1**  Sequence of control in a domestic premises: (a) line diagram; (b) circuit diagram

should check to see if such access equipment is safe to use and report any defects in a register. All site operatives should read *HSE Guidance Note GS 31: Safe use of ladders, step ladders and trestles*.

In (iii), a first-aid box should be provided by the employer and be accessible at all material working time. The first-aid box should be regularly checked for its contents. This topic is dealt with more fully in the author's *Part I Studies: Theory*, where reference is made to the *Health and Safety (First Aid) Regulations*.

In (iv), on a large project, an architect will insist on regular site meetings with the main building contractor and subcontractors. These meetings will be conducted with formal agendas, requiring minutes to be written on topics of importance, such as: site problems, variations, progress of work, visits etc.

7. For small projects, such as domestic dwellings, shops, offices, garages etc., you are more likely to meet the customer (or authorised visitor) than when working in a large project. In such situations, you represent your company's interest and your attitude and behaviour,

knowledge and standard of work are most likely to reflect or influence any future work that your company might obtain. It is important for you to keep regular time keeping, be tidily dressed, and adopt a friendly approach to the customer. This means being polite and respecting customers from different cultural background.

Where the customer is resident in the premises and the premises are furnished or at a stage of completion, take extra care of property and possessions. When drilling holes and making good, use a dust sheet and keep the working area clean and tidy. Any electrical work left unfinished should be made safe and the customer notified.

In terms of being asked questions by the customer (or an authorised visitor on site), you should be knowledgeable of your company's details, manager/supervisor names and your company's telephone number. You should be fully versed with the work you are doing and have knowledge of safety requirements.

On large sites, when working with other operatives, it is quite easy to get carried away with high spirits. Tomfoolery and pranks can be blamed for a large number of accidents and they only result in other people being distracted from essential tasks to sort out the problem. In such situations, you should exercise self-discipline and be aware of where you are and consider the cost of acting foolishly which could lead to suspension and possible dismissal from work.

8. (i) This is a document completed by a site operative informing the employer of the weekly time spent on a job together with any travelling time, mileage/fares allowance that may be due. It is verified and signed by the site foreman and sent back to the main office for payment. The office can use operative time sheets to give an assessment of the original estimate for the contract.

   (ii) This is a document (or voucher) that requires the recording of all time and materials used for a particular repair or alteration to a contract. It requires a job reference number and a description of the work and time taken to complete. For small projects the daywork sheet can be sent back to the main office but on large

projects it needs to be signed and approved by a contract administrator.

**Note**: A clause in a contract, under 'Schedule of Daywork' might state the following:

### Schedule of daywork

Additional work unless otherwise agreed shall be valued at daywork rates . . . and priced by the Electrical Contractor when tendering. Time sheets must be kept for inspection by the Contract Administrator, indicating the operatives' daily time spent on dayworks, together with the materials and plant used to execute the work. All time sheets are to be signed by operatives executing the work and the Electrical Contractor will be required to produce invoices for inspection of all materials used in connection with dayworks when submitting his final account.

9. (i) The architect is employed by the client to design the building. He is regarded as the leader of the building team and has to see that the building stays within the budget. He also recommends a suitable builder, controls the work and advises the client on progress.

   (ii) The clerk of works is the client's representative on site and reports to the architect. He offers advice to the main builder and liaises with the architect when problems occur.

   (iii) The quantity surveyor is chosen by the client on advice from the architect. He will advise the design team on approximate costs. He prepares bills of quantities and intermediate valuations of work done so that regular payments can be made to the main builder.

   (iv) The contracts manager is employed by the contractor to supervise the running of a number of different contracts. He has the overall responsibility for the planning of the building operations and is the link between the contractor's main office and the site agent or project manager under his control.

   (v) The structural engineer is employed by the client to design the structure of the building. He works closely with the architect and submits design calculations to the local authority for approval. As part of the design team (civil engineer and service engineers), he will make regular inspections of the work to ensure that the installation is carried out according to his design.

**10** See Regulation 12 of the *Electricity at Work Regulations* and Sections 476 and 537 of the *IEE Wiring Regulations*.

---

## EXERCISE 2.2

**1.** (i) Regulation 521–02–01 states that single-core armoured cables protected with steel wire or tape cannot be used on a.c. supplies. This is because the armouring is a ferrous metal and will become induced by the magnetic field around the cable conductor. This results in eddy currents circulating in the armouring and causing a rise in temperature at points between the gland termination and metal enclosure. These parts can get very hot and cause damage to a cable's insulation.

Regulation 523–05–01 allows single-core metal sheathed and/or non-magnetic armoured cables in sizes exceeding 50 mm² to be single point bonded (see Figure A4.3).

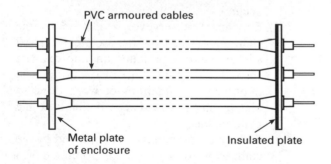

Figure A4.3   Single point bonding to stop the flow of eddy currents in single-core armoured cables

An alternative is to slot any ferrous metal enclosure at the point of termination.

(ii) Regulation 522–06–03 refers to the cables being insulated and having a metal sheath or armouring to provide mechanical protection. They may also be of the PVC insulated concentric type. The usual practice is to identify the position of the under ground cable with cable covers or yellow marking tape (see Figure A4.4).

The cables have to be at sufficient depth

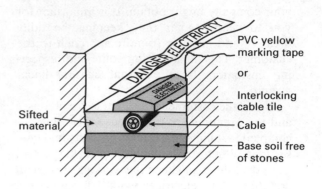

Figure A4.4   Precautions taken for a buried cable

to avoid damage from any disturbance. You should also refer to HSE GS 33 *Avoiding danger from buried electricity cables*.

(iii) Regulation 521–01–01 states that cables have to be installed out of reach of any source that may cause them damage. Refer to Section 5 and Table 13 of the *IEE Guidance Notes: Selection and Erection*. Cables should be bound with yellow and black tapes, in accordance with BS 2929 *Safety colours for use in industry* or alternatively, freely moving strips of fabric or plastics, attached to draw attention to the cables. Figure A4.5 shows one method used for wiring between buildings.

**Note**: For further reading on this topic, consult HSE GS 6 *Avoidance of danger from overhead electric lines* (Revised 1991)

**2.** (a)   See Figure A4.6.

**3.** See Regulation 522–06–05.
   (b)   See Figure A4.7.
   (c)   150 mm – See Regulation 471–13–04.

**4.** See Figure A4.8.

**5.** (a)   See Definitions and Section 528–01 of the *IEE Wiring Regulations*.
   (b)   See Figure A4.9.

**6.** Student Activity

**7.**      $I_b = 25$ A
      $I_n = 25$ A
      $I_t \geq 25/(0.725 \times 0.97) \geq 35.5$ A
Cable size initially chosen is 6 mm² taking 38 A.

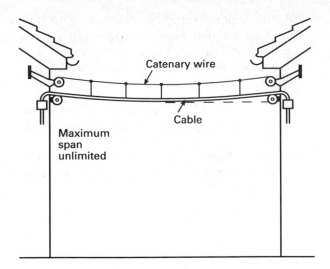

Figure A4.5   Wiring between building using a catenary wire to support wiring system (Appendix 4, IEE On-site Guide)

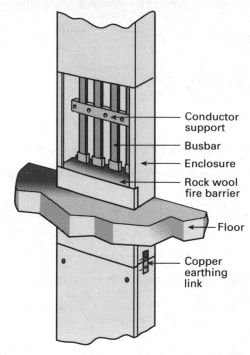

Figure A4.6   Busbar trunking in a rising main system

Voltage drop $V = 30 \times 25 \times 0.0073 = 5.5$ V (too high).
Choose and select 10 mm$^2$ cables with voltage drop of 3.3 V.

8.  (i)   See Wiring Regs Table 4B1
    (ii)  See *IEE On-site Guide* Table 4A
    (iii) See *IEE On-site Guide* Table 4E

9.  Answer is 200 mm × 200 mm.

10. Student activity.

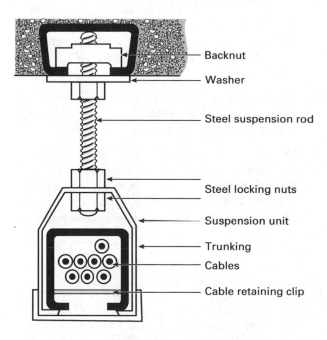

Figure A4.8   Trunking support bracket

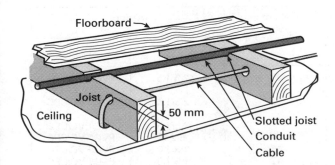

Figure A4.7   Methods of protecting PVC under floor boards

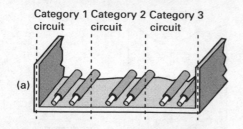

Category 1 Category 2 Category 3
circuit      circuit      circuit

(a)

Category 1 Category 2 Category 3 Category 3

(b)

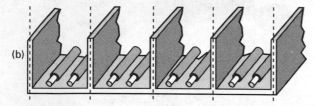

Figure A4.9   Category circuits in trunking: (a) common trunking (Cat. 2 insulated to Cat. 1 standard; Cat. 3 must be MI cable); (b) partition trunking; partitions to be fire resistant

## EXERCISE 3.2

1. See Figure A4.10.

2. (a)  See Reg. 514–12–02.
   (b)  See page 91, Chapter 4 of the author's *Electrical Installations Part 1 Studies: Theory*.

3. The answer to this question can be found in HSE, HS(G) 107 *Maintaining portable and transportable electrical equipment*, (1994).

4. The answer to this question can be found in *Electricians' Handbook – Guidance on BS 7671 1992 Requirements for Electrical Installations*, obtainable from the ECA. Page 130 of this guide lists such items as: client details, purpose of the report, occupier, description of the premises, estimated age, evidence of alterations and/or additions, records available, extent and limitations of the report, recommendations, non-compliance, etc.

5. (a)  The answer to this question can usually be found in product manufacturers' catalogues.
   (b)  See back page of the NICIEC's Emergency Lighting Inspection and Test Certificate.

6. The answer to this question can usually be found in a product manufacturer's catalogue on fire alarm systems.

7. The answer to this question can be found in HSE, HS(G)41 *Petrol filling stations: construction and operation*, 1990.

8. See page 96–97, Chapter 2 of the author's *Electrical Installation Technology*.

9. (a)  See HSE, GS 38 *Electrical test equipment for use by electricians*, 1991.
   (b)  See page 14 of the author's *Electrical Installation Part 2 Studies: Practical* book, 'Voltage measurement in a 3-phase system'.

| Circuit | | Protective device | | No of points | Wiring | |
|---|---|---|---|---|---|---|
| No | Description | Type | Rate (A) | | Size (mm²) | Type |
| 1 | Ring to downstairs sockets | Type 1 MCB RCD | 30 0.03 | 15 | 2.5/1.5 | PVC/PVC/CPC |
| 2 | Ring to upstairs sockets | Type1 MCB RCD | 30 | 10 | 2.5/1.5 | PVC/PVC/CPC |
| 3 | Water heater | Type1 MCB | 20 | 1 | 2.5/1.5 | PVC/PVC/CPC |
| 4 | Cooker circuit | Type 1 MCB | 30 | 1 | 6/2.5 | PVC/PVC/CPC |
| 5 | Upstairs lights | Type 1 MCB | 5 | 15 | 1.5/1.0 | PVC/PVC/CPC |
| 6 | Downstairs lights | Type 1 MCB | 5 | 12 | 1.5/1.0 | PVC/PVC/CPC |

Figure A4.10   Distribution board circuit schedule

## EXERCISE 4.2

1. (a) See definitions given in Part 2 of the *IEE Wiring Regulations*.
   (b) See Memorandum of the *Electricity at Work Regulations* or page 3 of the author's *Electrical Installation Technology 3: Advanced Work*.
   (c) See Figure 4.5.

2. (i) The Regulations stipulate that the space can only be accessible with the use of a tool. A side panel which can easily be removed by hand now needs to be fixed in position.
   (ii) A socket outlet is not permitted under Regulation 601–10–02 but a flex outlet point can be used. The switch controlling the towel rail can be placed outside the room or it can be controlled by a double-pole, pull-cord switch to BS 3676 (see Regulation 601–08–01).

3. Junction boxes and socket outlets are permissible. The degree of protection for enclosures is IP2X for indoor pools and IP4X for outdoor pools. Appliances may be Class I or Class II construction and with the exception of instantaneous water heaters to BS 3456. Equipment, switches, socket outlets, etc. have to be protected by one or more of the following: (i) electrical separation; (ii) SELV or (iii) a 30 mA r.c.d. A shaver socket outlet is allowed provided it complies with BS 3535. Socket outlets have to comply with environmental conditions and meet BS 4343.

4. It will be seen from Figure 4.10 that the stove is contained in Zone A and only equipment directly associated with this is permitted. Also, the wiring to the luminaire must be 150°C rubber insulation and be mechanically protected (see Reg 603–07–01). A switch outside the room would satisfy Regulation 603–08–01.

   The supply cable will pass through the room's thermal insulation and depending on its thickness, the size of the cable will be subject to a derating correction factor as appropriate in Table 52A of the *IEE Wiring Regulations*.

5. It is important that the earth electrode is sited in a position where it will not cause harm to the horses and it must be outside the resistance area of any other earth electrode such as an electric fence controller. All socket outlets should be protected by a 30 mA residual current device and give a maximum fault voltage of 25 V and out of reach of the horses.

6. The socket outlets must be provided with a SELV supply and have a voltage not exceeding 25 V. Their design must not be compatible with other types of socket outlet such as BS 1363 types used for 240 V supplies. Two locations are metal boilers, large metal containers that need cleaning, also some steel structures such as factory gantries.

7. (i) 0.75 mA, (ii) 0.75 mA/kW, and (iii) 0.25 mA.

8. It is important to note that if the supply was a TT earthing system and the socket outlets were locally protected by their own r.c.d's than their operation would have to discriminate with the supply r.c.d.

9. See definitions given in Part 2 of the *IEE Wiring Regulations*. In answer to (iv) street lighting may be classed as street furniture whereas a bus shelter that is lit or a telephone box that is lit is classified as street-located equipment.

10. See pages 76–77 of the author's *Electrical Installation Technology 3: Advanced Work*.

## Answers to multiple-choice questions chapter 5

| | | | | | | | | | |
|---|---|---|---|---|---|---|---|---|---|
| 01. | b | 21. | b | 41. | a | 61. | d | 81. | a |
| 02. | a | 22. | c | 42. | c | 62. | a | 82. | d |
| 03. | a | 23. | b | 43. | a | 63. | a | 83. | d |
| 04. | c | 24. | d | 44. | b | 64. | d | 84. | d |
| 05. | d | 25. | c | 45. | a | 65. | b | 85. | c |
| 06. | b | 26. | b | 46. | d | 66. | d | 86. | d |
| 07. | a | 27. | b | 47. | d | 67. | a | 87. | c |
| 08. | c | 28. | d | 48. | b | 68. | c | 88. | a |
| 09. | b | 29. | a | 49. | d | 69. | d | 89. | d |
| 10. | c | 30. | d | 50. | b | 70. | d | 90. | c |
| 11. | b | 31. | b | 51. | b | 71. | b | 91. | a |
| 12. | d | 32. | a | 52. | d | 72. | c | 92. | a |
| 13. | d | 33. | b | 53. | c | 73. | a | 93. | a |
| 14. | b | 34. | b | 54. | a | 74. | a | 94. | d |
| 15. | a | 35. | c | 55. | b | 75. | b | 95. | b |
| 16. | a | 36. | c | 56. | a | 76. | b | 96. | e |
| 17. | c | 37. | b | 57. | c | 77. | b | 97. | b |
| 18. | c | 38. | d | 58. | c | 78. | a | 98. | a |
| 19. | d | 39. | c | 59. | b | 79. | a | 99. | c |
| 20. | a | 40. | b | 60. | c | 80. | c | 100. | a |

## Answers to written questions

**Q1** See answers provided in Question 136 of the author's *Questions and Answers in Electrical Installation Technology*.

**Q2** (a) BS 951 earthing clamp to be used. Avoid corrosion of the clamp label in contact with dissimilar metal. Make proper mechanical connection between protective conductor and clamp. Check clamp connection on completion of work and label to be easily read.

(b) See Regulations 514–01–01, 514–03–01, 514–13–01, 543–01–01, 543–02–02, 543–02–03 and 543–03–03.

**Q3.** (a) See definition in Part 2 of the *IEE Wiring Regulations*.

(b) To encourage a rapid response of the circuit protective device when an earth leakage current flows.

(c) See Table 41A, *IEE Wiring Regulations*. Answer is 0.2 s

(d) See definition of both terms found in Part 2 of the *IEE Wiring Regulations*.

(e) 21 Ω.

**Q4** (a) 0.4 s (see Regulation 471–08-03 of the Wiring Regulations.

(b) 450 A

(c) 0.63 Ω

(d) 0.35 Ω

**Q5.** (a) See Q110 in the author's *Question and Answers in Electrical Installation Technology*.

(b) Table 2 of the *IEE Guidance Notes No. l: Selection and Testing* should be consulted.

(c) See Figure A4.11 (Follow the procedure given by the time/current characteristics in Appendix 3 of the *IEE Wiring Regulations*.

**Q6.** (a) The neutral is a live conductor and borrowing one from an independent circuit source (another circuit on a different fuse) could lead to electric shock. This is a possibility when neutrals are being disconnected and the independent circuit is switched on.

(b) See Regulation 314-01-04.

**Q7.** The answers to this question can be found in the author's *Electrical Installation Part 2 Studies: Practical*, Chapters 2 and 3.

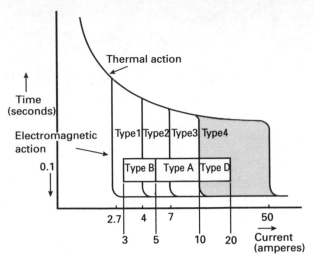

Figure A4.11   Characteristics of different types of miniature circuit breaker

**Q8.**  (a)  See BS 5266 or product manufacturer's emergency lighting catalogues (Menvier, JSB, Gents, etc.).

(b)  See BS 5839 or product manufacturer's fire alarm catalogues.

**Q9.**  See notes in Chapter 3.

**Q10.**  Some suggestions are as follows:

(a)  If the motor does not attempt to start, check terminal voltage with a proprietary tester; check voltage at distribution board and starter; check overload settings in the starter; check winding continuity and insulation resistance.

(b)  If motor attempts to start, check movement of the rotor/mechanical loading; check setting of overloads; check mechanical load specifications.

**Q11.**  See page 41, Chapter 3 of the author's *Electrical Installation Technology 3 Advance Work*.

**Q12.**  See information contained in HSE, GS 6 *Avoidance of danger from overhead electric lines*.

**Q13.**  See Chapter 3.

**Q14.**  (1)  Legislation – the written or enacted law in the form of statutes, acts etc., deriving its authority directly or indirectly from Parliament.

(2)  Civil law – the law related to the rights, duties and obligations of people, property, contracts, insurance, etc.

(3)  Criminal law – the law dealing with wrongful acts harmful to the community and punishable by the State.

(4)  European Community law – the additional source of law by our accessation treaty and enactment of the European Communities Act of 1972 (this takes the form of European directives).

(5)  Statutory instrument – the subordinate legislation made by government departments.

**Q15.**  Student activity.

**Q16.**  (1)  PVC/PVC sheathed cable routes: to be straight, vertical, horizontal and parallel to walls unless shown otherwise; to be run parallel and adjacent to heating pipes and positioned at least 150 mm clear of other services; to be run horizontally and concealed in walls and to be located within 150 mm of the ceiling or between 150 mm and 300 mm of the floor; to be concealed and where run to switches and outlets, they are to be vertically in line with the accessory.

(2)  and (3)  Follow the same procedure as (1).

**Q17.**  Student activity.

**Q18.**  Some hints are given on page 86, Appendix 1 of the author's *Electrical Installations Part 1 Studies: Practical*.

**Q19.**  See other books in the *Electrical Installations Competences Series*, also a firm's safety policy.

**Q20.**  (a)  See procedure on page 58, Chapter 3 of the author's *Electrical Installation Part 2 Studies: Practical*.

(b)  See Figure A4.12.

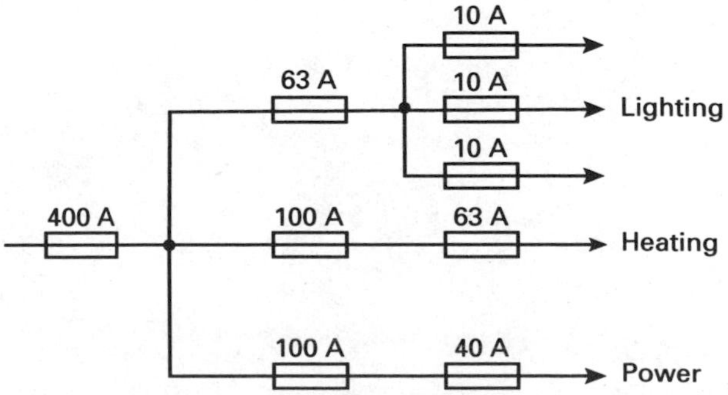

Figure A4.12   Fuse discrimination

# Index

programme of work, 17, 25
protection for safety, 65
provisional sum, 18

regulations,
   - Building, 1, 7
   - Construction (Head) Protection, 1, 5
   - COSHH, 1–4
   - Electricity at Work, 1, 4, 6ff
   - Electricity Supply, 1, 2, 80
   - Employment Protection (Consolidation) Act, 18
   - HSAW Act 1, 2, 5, 14ff
   - IEE Wiring Regulations, 1, 6, 7, 15ff
   - Low Voltage, 1, 6

reports, 27
residual current device, 66, 71–78
restrictive conductor locations, 76
retention, 18

schedule of rates, 18
schedules, 15, 23

SELV, 65ff
site
   - diary, 26
   - office, 23
   - plan, 8
specifications, 8, 13–15, 23, 52, 56
statutory instrument, 2
swimming pools, 70–71

technician, 20
tendering, 8, 15–17
testing and measurement, 54
time sheet, 26–27
trunking, 40–41
TUC, 2

variations, 17
verification, 52
visual inspection, 53
voltage drop, 45

wiring systems, 32